ESSENTIALS OF MULTIVARIATE DATA ANALYSIS

ESSENTIALS OF MULTIVARIATE DATA ANALYSIS

NEIL H. SPENCER

CRC Press
Taylor & Francis Group
Boca Raton London New York

CRC Press is an imprint of the
Taylor & Francis Group, an **informa** business

A CHAPMAN & HALL BOOK

CRC Press
Taylor & Francis Group
6000 Broken Sound Parkway NW, Suite 300
Boca Raton, FL 33487-2742

© 2014 by Taylor & Francis Group, LLC
CRC Press is an imprint of Taylor & Francis Group, an Informa business

---

Library of Congress Cataloging-in-Publication Data

Spencer, Neil Hardy.
　　Essentials of multivariate data analysis / Neil H. Spencer.
　　　　pages cm
　　"A CRC title."
　　Includes bibliographical references and index.
　　ISBN 978-1-4665-8478-5 (hardcover : alk. paper)
　　1. Multivariate analysis. I. Title.

QA278.S683 2014
519.5'35--dc23
2013045984

Visit the Taylor & Francis Web site at
http://www.taylorandfrancis.com

and the CRC Press Web site at
http://www.crcpress.com

For Catherine, Laura, Julia and Helen

Contents

Preface

Why does this book exist? Well, essentially because (1) multivariate statistical methods can be very useful to people doing applied research or learning how to do it (by this I mean exploring data and using it to answer research questions); (2) most other books on multivariate methods are aimed at statisticians or researchers who are comfortable with (and enjoy?) mathematics and formulae. The aim of this book is to explain the usefulness of multivariate methods in a way which is accessible to students and researchers who would not consider themselves statisticians or mathematicians. They may be put off by the whole idea of quantitative analysis of the formulae they see in other books. Here they will find very few formulae, and those that cannot be left out are made to seem less scary than they might look.

"But surely most researchers have been trained in statistics?" I hear you say. This is true. Students may also have had some statistics training before they find this book. However, their training may have concentrated on topics like summary statistics, graphical displays, confidence intervals, hypothesis testing, correlation and regression. Although multivariate data may have had a look when studying multiple regression, most of the statistical training will have considered one variable at a time rather than many variables together. An exception to this may be factor analysis which is widely used in the social sciences but the other topics discussed in this book are less likely to have been encountered. Most datasets are multivariate (containing a number of variables) and as a result, multivariate methods are useful to explore and to use them to answer research questions.

"But surely we have to use complicated computer packages to do multivariate analyses?" I hear you say. Well, not if you are trying to do straightforward analyses. Accompanying this book is an add-in for Microsoft® Excel® which can be used to do the analyses shown here. Sure, if you are looking to do more complicated things, then you will need to move to a specialist package such as Minitab®, R, SAS®, SPSS®, Stata®, etc. However, if you have already learnt the essentials from this book, then the transition will not be too hard and you may find you do not need to get involved with these at all.

"But surely there are other multivariate statistical methods which you have not included in your book?" I hear you say. Absolutely true. There are lots of methods out there but this book concentrates on the most commonly used ones; it is titled *Essentials of Multivariate Data Analysis* rather than *A Comprehensive Guide to Multivariate Data Analysis*. Sticking to the essentials is what makes this book useful and, although it is a good introductory read for students studying more mathematical/statistical courses, there are already plenty of books which will give them the depth and breadth they want. A number are suggested in the "More Information" sections at the end of each topic.

Dr Neil H Spencer
University of Hertfordshire

Frequently Asked Questions

1

1.1 WHAT QUESTIONS?

I start the book by trying to answer two frequently asked questions: (1) What analysis should I do? and (2) What data do I need?. These two questions are of course very closely linked to each other with the answer to the "What analysis should I do?" restricting the range of possible answers for "What data do I need?".

Of course, before this, I hope that you have a 'research hypothesis' of some kind. This sounds rather grand, but what it really means is, 'What are you trying to find out about?'. This should be the starting point for any study, and the answer to this question will lead to the answer to 'What analysis should I do?'.

Next in this chapter is the answer to the question: What data is the author using in this book? Perhaps this is not a frequently asked question but I hope the reader will forgive me for this. Certainly if I do not explain the dataset somewhere, then 'What data is the author using in this book?' would become a frequently asked question from readers.

I conclude this chapter by answering a few questions about missing data, the choice of topics for this book and the issue of computer packages.

1.2 WHAT ANALYSIS SHOULD I USE?

You will not be surprised that in a book I cannot answer this question completely for your particular research. What I am able to do, however, is give some outline examples of the sorts of things you might be wanting to do and point you in the direction of the multivariate techniques that are contained in this book. In this way, this section is more like the American television quiz show *Jeopardy,* in which the contestants are given answers and have to guess the questions. In this book, the following chapters give the answers and here I am suggesting questions to which they are the answer.

At this point I should also say, 'Please consult a Statistician'. I am saying this, not because I am trying to make Statisticians such as myself seem important and clever people (which we are!) or create jobs for Statisticians (described in 2009 as "the sexy job in the next 10 years" by Google's Chief Economist, Hal Valerian), but because a Statistician is (hopefully!) going to have a better idea of whether you are choosing the correct analysis than you will, even after you have read this section of the book. At the very least a consultation with a Statistician will give you confidence that the analysis you are embarking upon is able to answer the questions you want answered. The consultation may even point you in new directions which are even more beneficial to your work.

- *I want to look at the data to detect patterns, groups and unusual cases. What analysis should I use?* If your analysis aims to explore the data, then have a look at Chapter 2: "Graphical Presentation of Multivariate Data". You may also want to think about non-multivariate numerical summaries, looking at one variable on its own at a time (univariate analyses) or the relationship between pairs of variables (bivariate analyses).

- *I want to check if differences between groups across a number of variables in terms of averages or the amount of variation are real or just random fluctuations. What analysis should I use?* If your analysis aims to test hypotheses about whether groups are the same or different from each other, then have a look at Chapter 3: "Multivariate Tests of Significance". You may also be interested in seeing if there are differences between groups in terms of just one variable at a time. For this, traditional univariate hypothesis tests (not discussed in this book but covered in innumerable others) can be used.

- *I have a number of variables and want to understand the underlying causes of the responses I have – what factors are at work that come together to bring about the observed data. What analysis should I use?* If your analysis aims to explore the data and try and discover underlying reasons for the data being as they are, then have a look at Chapter 4: "Factor Analysis", thinking in terms of undertaking exploratory factor analysis. You may also want to examine the available literature to see if there are any theories applicable to your data that might suggest particular factors exist. If so, then you might want to undertake the factor analysis from a confirmatory point of view.

- *I have a number of variables and want to test a theory about the underlying causes of the responses I have – are the factors that come together to bring about the observed data what the theory suggests? What analysis should I use?* If your analysis aims to try and confirm that particular underlying reasons exist for the data being as they are (or suggest that the theory is wrong), then have a look at Chapter 4: "Factor Analysis", thinking in terms of undertaking confirmatory factor analysis. You may also want to look at the topic of Structural Equation Modelling (SEM), particularly if you want to test a theory which involves a complex relationship between the underlying factors and observed variables. This is an advanced topic which is not discussed in this book. For a good introduction, I recommend Bartholomew et al. (2008).
- *I want to understand my data better by discovering any groups that might exist and looking at their characteristics. What analysis should I use?* If your analysis aims to find groups which may exist in your data and examine them, then have a look at Chapter 5: "Cluster Analysis". You may also want to initially undertake some graphical analyses which may reveal possible clusters (see Chapter 2: "Graphical Presentation of Multivariate Data").
- *I have information in my data which tells me the cases are in particular groups and I want to be able to allocate any new cases to one of these groups based on particular characteristics which have been recorded. What analysis should I use?* If your analysis aims to work out which of a number of groups is the most likely for any new cases to belong to, then have a look at Chapter 6: "Discriminant Analysis". You may also want to look at logistic regression (if you have two groups) or multinomial regression (if you have more than two groups). I am not discussing regression-based techniques in this book but, for a good introduction, I recommend Field (2009).
- *I have information in my data which tells me the cases are in particular groups and I want to understand how being in one group or another is related to values of particular variables in the dataset. What analysis should I use?* If your analysis aims to find out whether a case's data for particular variables are related to the chances of being in one group or another, then have a look at Chapter 6: "Discriminant Analysis". You may also want to look at model building techniques in relation to logistic regression (if you have two groups) or multinomial regression (if you have more than two groups). As mentioned above, I am not discussing

regression-based techniques in this book but recommend Field (2009) for some introductory reading on these topics.

- *I have data which are measures of similarity/dissimilarity (or distances) and I want to have an overall graphical view of how similar or dissimilar the cases are. What analysis should I use?* If your analysis aims to plot the cases in a two-dimensional scatterplot (or more than two dimensions might just be feasible) in such a way that the measures of similarity/dissimilarity in the data are represented on the plot as closely as possible, then have a look at Chapter 7: "Multidimensional Scaling". You may also want to examine the scaled eigenvectors which are used to create the plot. They may be able to give you additional insight into any underlying factors which are bringing about the observed measures of similarity/dissimilarity.

- *I have data which are measures of similarity/dissimilarity (or distances) and I want to understand the underlying causes of these measures. What analysis should I use?* If your analysis aims to try and discover underlying reasons for the data being as they are, then have a look at Chapter 7: "Multidimensional Scaling". You may also want to consider producing plots of the scaled eigenvectors that you will be examining in order to gain a better understanding of which cases are similar or dissimilar.

- *I have a number of categorical variables and want to explore the relationships that exist between the different categories of the different variables. What analysis should I do?* If you want a graphical representation of the relationships between the different variables and their categories, then have a look at Chapter 8: "Correspondence Analysis". You may also want to look at log-linear modelling which aims to build a model to explain the data. As mentioned a couple of times above, I am not discussing regression-based techniques in this book but, again, a good introduction to this topic can be found in Field (2009).

1.3 WHAT DATA DO I NEED?

As the answer to this question depends largely on the answer to 'What analysis should I do?', you will find in each of the following chapters a section entitled "What Data Do I Need for *[topic]*?". However, many of the multivariate techniques in this book require continuous data, or data which can be treated as continuous. This is any data for which calculating a mean or a standard deviation is sensible. Discussing the mean or standard deviation of people's hair colour makes no sense, so data of this type are not continuous (and are sometimes called *nominal*: the data represent categories, and these categories have no meaningful order in which they can be put).

I should mention binary data at this point. This is categorical data for which there are just two categories, such as "male" and "female". In this case, if "male" is coded 1 and "female" is coded 0, then the mean does have an interpretation: it is the proportion of the sample that is male. However, this does not make it continuous data. If a coding system other than 0/1 had been used, the mean would not be interpretable. However, having said this, some of the techniques discussed in this book can use this binary data as if it were continuous.

Some categorical data can be treated as if it were continuous data, providing care is taken. If the categories can be placed into a natural order, then scores can be given to each category and these scores treated as continuous variables. For instance, take the example of data collected from a survey where respondents have been asked to state how satisfied they are with a service they have purchased. They may be asked to put themselves into one of the following categories: "very satisfied", "satisfied", "neither satisfied nor dissatisfied", "dissatisfied" or "very dissatisfied". This kind of question is sometimes referred to as a Likert scale question. What we have here is a categorical variable where the categories have a natural order to them. We could then proceed to give scores to these categories, so "very satisfied" might score 10, "satisfied" might score 8, "neither satisfied nor dissatisfied" might score 5, "dissatisfied" might be given a score of 2 and "very dissatisfied" might score 0. We could then treat these scores as if they were a continuous variable.

However, before you go off and launch into giving scores to any Likert scale data you possess, let me sound a note of caution. The scoring system I have constructed above is just one of many scoring systems that I could have used. Hopefully when you give scores to your Likert scale questions, you will be able to use some expert knowledge about the situation from which your data have been collected, and devise a scoring system that can be defended if necessary.

Please (and I beg of you) do not do the following. Frequently, Likert scale responses are coded so that, for instance, 1 indicates "very satisfied", 2 indicates "satisfied", 3 is the code for "neither satisfied nor dissatisfied", 4 means "dissatisfied" and 5 means "very dissatisfied". This coding is normally done to make data entry easier and because statistical analysis packages prefer to operate with numbers as data. Time and time again, you will see analyses conducted where this 1 to 5 coding is used as the scoring system. This implies that, for example, moving from "satisfied" to "very satisfied" (an upward change of one unit in this scoring system) is the same as moving from "dissatisfied" to "neither satisfied nor dissatisfied". Now, if you are happy that for your Likert scale this is an appropriate scoring system to use, then by all means go ahead and use it. But please do think about whether it is an appropriate scoring system and do not simply use the 1 to 5 coding system as a scoring system because it is already there.

1.4 WHAT DATA IS THE AUTHOR USING IN THIS BOOK?

The dataset being used in this book is entirely fictitious. I did not decide to do this because I did not want the data to be realistic, but rather because I need it to be relatively straightforward and have useful features which help explain the multivariate techniques. It needs to be straightforward because if I choose a dataset which is unusual or difficult to understand, then your attention will be distracted from the multivariate techniques which are (I am assuming!) what you want to understand. It needs to have useful features which help explain the multivariate techniques because there is little point in trying to explain, for instance, cluster analysis with a dataset which does not contain any clusters.

Although I fully admit that the data are fictitious, I have nevertheless made efforts to ensure that they are realistic. The data have been generated with means, standard deviations and correlations that are similar to those found in the UK adult population. The data can be downloaded from the publisher's website at http://www.crcpress.com/product/isbn/9781466584785.

The dataset consists of 100 people for whom we have created the following data:

- Gender: 48 are male, 52 are female;
- Age: the youngest is 18 and the oldest is 69 – we have this data both in terms of age in years and age-groups 18–29, 30–39, 40–49, 50–59, 60+;
- Height and weight: in metres and kilogrammes, respectively;
- Blood pressure: systolic and diastolic, in mmHg;
- Pulse rate at rest: in beats per minute;
- Smoking history categorised as "never smoked", "occasional smoker", "ex-smoker" and "current smoker";
- General knowledge scores in nineteen subject areas (see below for more details); and
- Opinions about similarity of seven nations' foreign policies (see below for more details).

1.4.1 General Knowledge Scores

The general knowledge scores were created for nineteen subject areas, as below. These areas have been used in Irwing et al. (2001) and Lynn et al. (2001) to examine gender differences in general knowledge. Mean scores and standard deviations for males and females from Lynn et al. (2001) and correlations between variables from Irwing et al. (2001) have been used in the creation of the scores for the 100 cases in the dataset.

1. History of science	8. General science	14. Biology
2. Politics	9. Geography	15. Film
3. Sports	10. Cookery	16. Fashion
4. History	11. Medicine	17. Finance
5. Classical music	12. Games	18. Popular music
6. Art	13. Discovery and	19. Jazz and blues
7. Literature	exploration	

1.4.2 Opinions about Similarity of Nations' Foreign Policies

We imagine that the 100 people in the dataset are asked to judge how similar they believe the foreign policies of different countries to be. We have seven countries: United Kingdom (U.K.), United States (U.S.A.), France, Germany, Russia, China and Australia. For each pair of countries, each person has assigned a score to how similar they believe the countries' foreign policies to be, with 1 indicating very similar and 10 indicating very dissimilar. There are twenty-one pairs that can be created from the list of seven countries ($7 \times 6 \div 2 = 21$), and thus twenty-one variables.

1.5 WHAT ABOUT MISSING DATA?

Detailed information about how to deal with missing data is missing from this book, quite intentionally. The reason is this is a book about the essentials of multivariate data analysis and cannot cover everything. Of course you are almost bound to have some missing data in any real dataset you handle, particularly if human beings have been involved in any part of the process of creating it. We are clumsy, free-willed creatures who are bound to occasionally enter rubbish data ('Is this person really 234 years old?'), refuse to answer questions ('It is no concern of yours how much I earn – mind your own business') and use equipment which malfunctions ("This person has a blood pressure reading of zero – call an ambulance!"). The default position of any statistical analysis is to say, 'This line of the dataset does not contain all the information needed so we cannot use this line at all'. It's not a brilliant strategy because you are throwing away perfectly good information just because a little bit is missing. There are methods that exist to cope better with missing data but they are beyond the scope of this book. In any event, the best advice that can be given is to try and do all you can to avoid missing data in the first place. You can use data entry procedures that prevent impossible ages being entered; you can design questionnaires that explain why the research needs to know about income and will protect the information; you can make sure that equipment malfunctions are spotted at once and measurements retaken.

1.6 WHAT ABOUT OTHER TOPICS?

The topics included in this book are ones that I consider to be the most essential topics in multivariate analysis. There are numerous other topics which could come under the heading of "multivariate analysis" but are not included because they are less commonly used. This book is not trying to be an exhaustive guide to the subject area and, although there are bound to be some people who disagree with my choice, I am convinced that I have included all the most important topics.

1.7 WHAT ABOUT COMPUTER PACKAGES?

Accompanying this book is an add-in for Microsoft® Excel® which can be used to carry out the analyses shown in the different chapters. Some people will be horrified at the idea of Excel being used to do any sort of statistical analysis, let alone multivariate analysis. Excel is not a statistics package and has been shown to produce some inaccuracies in certain situations. An Internet search will find a number of websites and papers which point out some of the flaws and inconsistencies. However, just because it is not perfect does not mean it is useless. The add-in accompanying this book uses Excel mainly as a rather extensive calculator rather than relying solely on its in-built statistical functions. I may not be completely confident in its performance when it comes to the eighth decimal place in some calculations, but we should not be worrying about or reporting that level of accuracy. As you will find as you read this book, many multivariate analyses include a good amount of subjective decision making and interpretation by the researcher, and the numbers are there as a guide to help this process. The eighth decimal place should not really matter.

Having said this, it is undoubtedly the case that if you know how to use it properly, dedicated statistical software is superior to Excel and to the Excel add-in. These packages (e.g. Minitab®, R, SAS®, SPSS®, Stata®) are not only superior when it comes to the precision of their calculations, but also have the capacity to go beyond the basic analyses available in the Excel add-in and undertake more refined analyses. Hopefully, having understood the material in this book and the results from the Excel add-in, readers who need to take this route will be in a good position to get a grip on whichever package they choose to use.

Graphical Presentation of Multivariate Data

2

2.1 WHY DO I WANT TO DO GRAPHICAL PRESENTATIONS OF MULTIVARIATE DATA?

Before conducting any data analysis, it is always a good idea to look at the data first. No matter how well you feel you know the data, there is always the possibility that it contains some unexpected characteristic. One very good reason for doing this data examination before any analysis is that the results of looking at the data may cause you to change the dataset. If you look and see an unusual case in the dataset (perhaps it has very high or low values for one or more variables), then you may want to check it out in some way. Perhaps there has been a problem in data entry and the case in question has had its data input incorrectly. For instance, in a study involving children, one would be surprised to find a 99-year-old. However, closer inspection may reveal that the child was in fact 9 years old and the 99 recorded is due to a slip of the finger when inputting the data. Alternatively, you may discover that you have cases in your dataset that should not be there. If you are conducting a study involving children and discover a genuine 99-year-old in the dataset (that is, the 99 is correct and not a data input error), then you will probably want to remove him or her because they do not belong to the population you want to study.

Another reason for looking at the data is to try and spot patterns and relationships between variables. Perhaps there are groups of cases in your data that you had not suspected. The discovery of such groups may affect the hypotheses you then examine and the nature of the analyses you undertake.

So how do you actually go about looking at the data? This is easy to do in a univariate situation. There are plots like a histogram, bar chart, box plot, and the like which will give a visual display of a single variable. Summary statistics such as the mean, standard deviation, minimum, maximum, median, and upper and lower quartiles can also be easily calculated.

We can also easily move outside the univariate framework to a bivariate situation where we examine two variables at once. Scatterplots can show the relationships between the variables and summary statistics such as correlation can be calculated. It is even possible to create scatterplots of three variables and thus simulate a three-dimensional representation. However, care must be taken with these because on a computer screen or a printed page, one only ever sees one two-dimensional image at any time. Clever computer animations can be used to rotate an image on a screen to give it a three-dimensional appearance, and this may help interpretation. It is also sometimes the case that a three-dimensional plot can hide characteristics as well as display them (e.g. one bar on a chart may be hidden behind another). It is also possible to expand upon a two-dimensional scatterplot by adding a third variable which governs the size of the marker displayed on the plot. These are sometimes called "bubble" plots because the third variable makes the markers look like different sized "bubbles". You can even add a fourth dimension to these bubble plots if your dataset includes time and has animations which show how things are changing as the days/weeks/months/years go by. However, although these graphs can be very informative and even beautiful at times, they are not truly multivariate because they are still restricted to a relatively small number of variables.

This is a book about multivariate methods so you might think that I will simply ignore these univariate and bivariate techniques and refer you to other books. Well, while I am indeed going to send you in the direction of other books for the details of these techniques, I am not going to ignore them altogether. For the graphical presentation of multivariate data, it can be useful to combine a number of univariate and bivariate graphs, as you will see below. However, if you want to read more about the non-multivariate use of these univariate and bivariate methods, do look at books such as Field (2009).

2.2 WHAT DATA DO I NEED FOR GRAPHICAL PRESENTATIONS OF MULTIVARIATE DATA?

I suppose the answer to this is, 'It depends'. For all the methods described here, continuous data (or data that can be treated as continuous) are preferred. See Chapter 1 for a discussion of types of data. For the star plots and Chernoff faces, it is possible to use categorical data, although their usefulness is compromised slightly by this. For categorical data, we need a whole topic in its own right – see Chapter 8 on correspondence analysis.

2.3 THE REST OF THIS CHAPTER

I present here seven different methods of graphical display for multivariate data – three which are based on using univariate and bivariate techniques, three which are genuinely multivariate in their own right and one which is multivariate but does not display all the information in the dataset. However, before going any further, I should say that there is no really good way of displaying multivariate data. In our three-dimensional world, the best we can hope for in a graphical display is a three-dimensional model. In practical terms, we are most often restricted to a two-dimensional display on a computer screen or on paper. The methods described below are thus ways of overcoming our universe's physical limitations, and are therefore limited in what they are able to achieve.

The first three methods (comparable histograms, multiple box plots and the trellis plots – the ones that are based on univariate and bivariate graphs) are all good for comparing variables and the relationship between pairs of variables. However, they are not very helpful if you want to be able to see how individual cases in the dataset vary across the different variables in the dataset. Of the three truly multivariate methods, star plots and Chernoff faces struggle when there are even moderate numbers of cases in the dataset but can display a small number of variables with some degree of clarity. Andrews' plots cope better with larger datasets but the meaning of individual variables is lost to a large degree. The last method, a principal components plot, suffers from not being able to display all the information in the data. Sometimes, quite a lot of information can be missing but on other occasions, the plot can include almost all the information available in the data. If that sounds a bit vague, then do not worry because it is explained further in Section 2.14.

2.4 COMPARABLE HISTOGRAMS

You may feel a bit cheated by this method of graphical presentation. All we do is create a histogram for each of the variables and look at them all to spot patterns, outliers, etc. What is multivariate about this, and why the title "Comparable Histograms"? Well, the trick used here that makes it multivariate is to make sure that the horizontal and vertical axes on all the different histograms are the same. From the dataset discussed in Chapter 1, Figure 2.1 shows comparable histograms for the four variables history of science score, politics score, sport score and history score. Because the horizontal and vertical axes are the same in all the univariate graphs that make up Figure 2.1, we can clearly see that the sport score and politics score are more spread out than the history of science score and the history score. If we were to simply use the default settings for the horizontal and vertical axes for each of the histograms, we would get Figure 2.2. This makes the variables look like they all have very similar distributions.

2.5 A STEP-BY-STEP GUIDE TO OBTAINING COMPARABLE HISTOGRAMS USING THE EXCEL ADD-IN

1. You must have a column in Excel that contains the names by which your cases are known. These are called the "case identifiers". They may be names or codes that you can use to identify the different cases, or may be simply case numbers (e.g. case 1, case 2, etc.). You must also have columns of data containing the variables for which you want to create comparable histograms.
2. Go through the multivariate analysis add-in's menus until you get the dialogue box for comparable histograms.
3. In the "Case identifiers:" (e.g. names) box, put the range of cells corresponding to the column in which the case names, labels or whatever (see Step 1) are located.
4. In the "Variables for histograms:" box, put the range of cells corresponding to the columns that contain the variables for which you want to create the histograms.

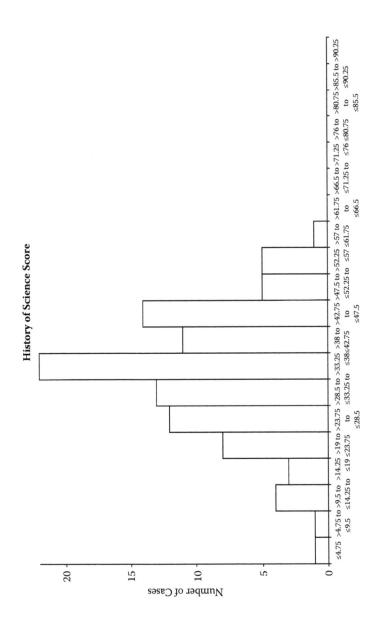

FIGURE 2.1   Comparable histograms.

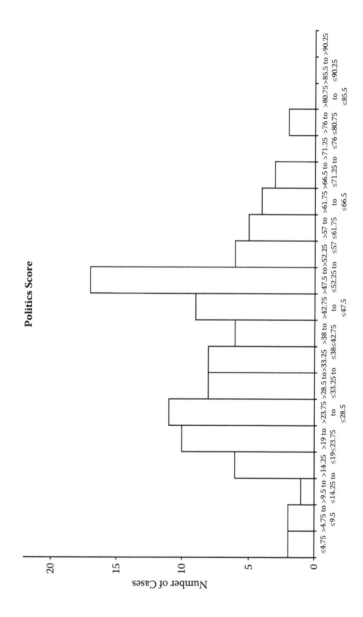

FIGURE 2.1 (Continued)

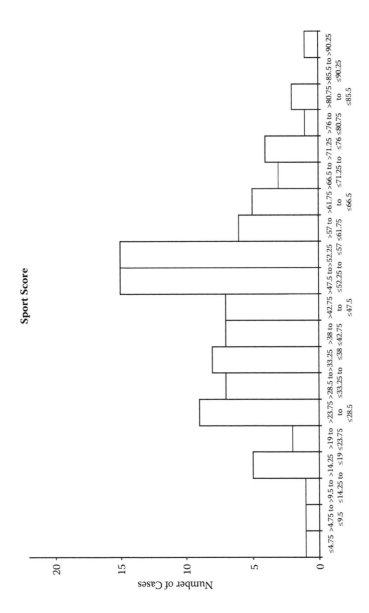

FIGURE 2.1 (Continued)

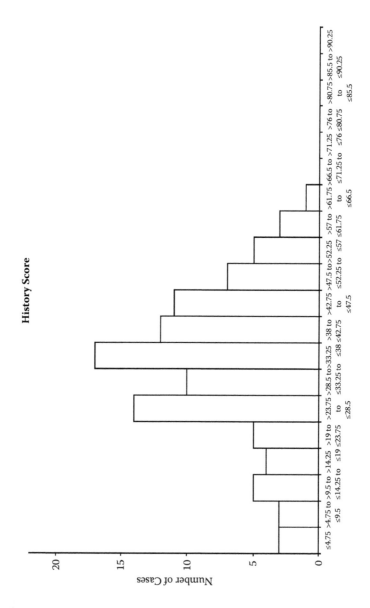

FIGURE 2.1 (Continued)

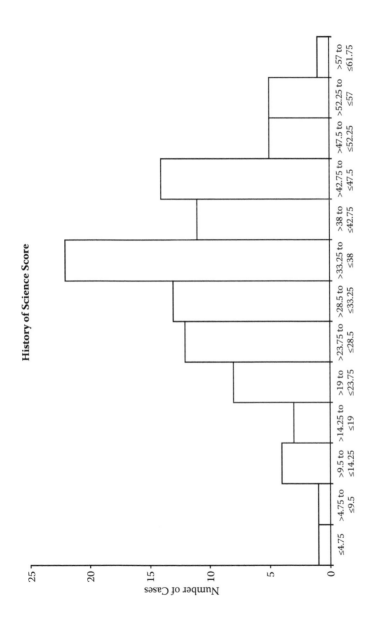

FIGURE 2.2 Non-comparable histograms.

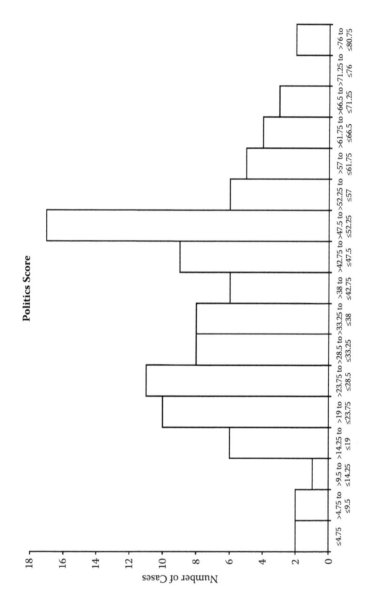

FIGURE 2.2 (Continued)

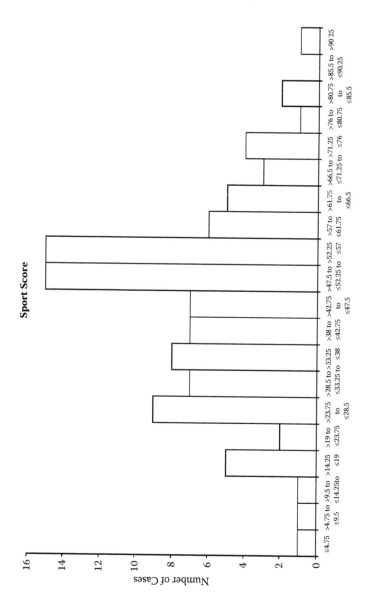

FIGURE 2.2 (Continued)

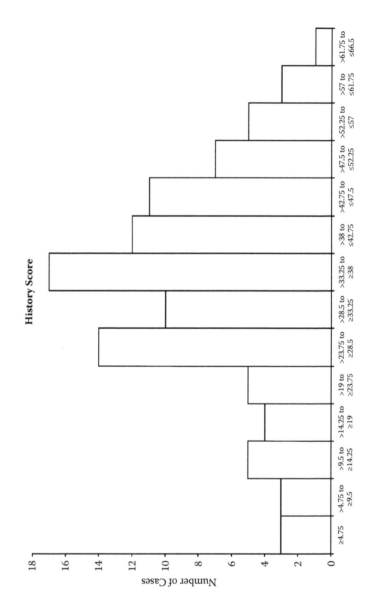

FIGURE 2.2 (Continued)

5. Make sure the choice for "Titles in the first row" or "No titles in the first row" is appropriate for the ranges you have entered at Steps 3 and 4.
6. Click "OK".

The comparable histograms should now be created and shown in a new workbook in Excel.

2.6 MULTIPLE BOX PLOTS

Like the comparable histograms, you may think that multiple box plots is a very straightforward idea. All we do is create a separate box plot for each of the variables in which we are interested and plot them on one graph. However, although it is not rocket science, it is easy to forget how useful these box plots can be – in one graph you can have information on a number of variables presented in such a way that you can compare the variables easily as well as see the distribution of each individual variable.

From the dataset discussed in Chapter 1, Figure 2.3 shows multiple box plots for the four variables history of science score, politics score, sport score

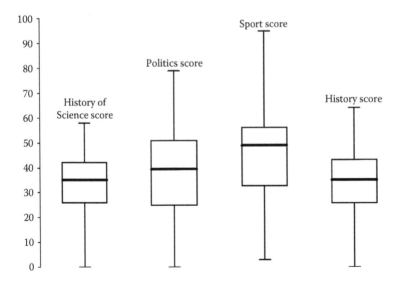

FIGURE 2.3 Multiple box plots.

and history score. As with normal univariate box plots, the thick lines across the middle of the boxes indicate the position of the medians, with the tops of the boxes being the upper quartiles (75% of the way through the data) and the bottoms of the boxes being the lower quartiles (25% of the way through the data). Here we have "whiskers" extending to the minimum and maximum values for each of the variables. You will sometimes see computer packages with shorter whiskers, and "outliers" and "extreme values" indicated by markers beyond the whiskers. The definitions of an "outlier" and what is "extreme" are, however, arbitrary ones and we thus prefer to have our whiskers cover the entire range of variables.

2.7 A STEP-BY-STEP GUIDE TO OBTAINING MULTIPLE BOX PLOTS USING THE EXCEL ADD-IN

1. You must have a column in Excel that contains the names by which your cases are known. These are called the "case identifiers". They may be names or codes that you can use to identify the different cases, or they may be simply case numbers (e.g. case 1, case 2, etc.). You must also have columns of data containing the variables for which you want to create comparable histograms.
2. Go through the multivariate analysis add-in's menus until you get the dialogue box for multiple box plots.
3. In the "Case identifiers:" (e.g. names) box, put the range of cells corresponding to the column in which the case names, labels or whatever (see Step 1) are located.
4. In the "Variables for box plots:" box, put the range of cells corresponding to the columns that contain the variables for which you want to create the box plots.
5. Make sure the choice for "Titles in the first row" or "No titles in the first row" is appropriate for the ranges you have entered at Steps 3 and 4.
6. Click "OK".

The multiple box plots should now be created and shown in a new workbook in Excel.

2.8 TRELLIS PLOT

If you have not heard of this type of plot before, you may be imagining that the name has something to do with gardening and plants climbing up and over a wooden trellis. I am sorry to disappoint but the connection has to do with how these garden trellises are usually constructed with horizontal and vertical pieces of wood making up a grid pattern. A trellis plot is made up of a number of univariate and bivariate graphs arranged in a grid in such a way that visual comparisons can be made. There is no one particular definition of what a trellis plot should contain but typically a scatterplot is created of each variable of interest against each of the other variables of interest.

Thus if we have four variables from the dataset discussed in Chapter 1 – history of science score, politics score, sport score and history score – we have $4 \times 3 = 12$ scatterplots. These can be arranged in a 4×4 grid (see Figure 2.4)

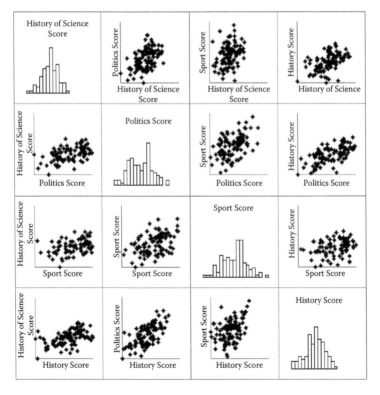

FIGURE 2.4   Trellis plot.

with the first variable (history of science score) being involved in all the plots in the first row and in the first column, the second variable (politics score) being involved in all the plots in the second row and in the second column, etc. For the top left part of the grid, we could show a scatterplot of history of science score against history of science score but that would be rather pointless. We could instead leave that cell of the grid blank but here we have included a comparable histogram (see Section 2.4) and have also done the same for the other parts of the grid on the diagonal from top left to bottom right.

You might complain that the graphs in the trellis plot do not have any scales or proper titles on them and that the axis labels are rather small. You would be correct in noting these shortcomings. However, the purpose of the trellis plot is to gain a visual understanding of the patterns and relationships between the variables and this can be done easily enough from Figure 2.4. If you want to go on to examine any of the graphs in more detail, then simply create larger versions of the ones you want.

2.9 A STEP-BY-STEP GUIDE TO OBTAINING A TRELLIS PLOT USING THE EXCEL ADD-IN

---

1. You must have a column in Excel which contains the names by which your cases are known. These are called the "case identifiers". They may be names or codes that you can use to identify the different cases, or may be simply case numbers (e.g. case 1, case 2, etc.). You must also have columns of data containing the variables for which you want to create comparable histograms.
2. Go through the multivariate analysis add-in's menus until you get the dialogue box for the trellis plot.
3. In the "Case identifiers:" (e.g. names), box, put the range of cells corresponding to the column in which the case names, labels or whatever (see Step 1) are located.
4. In the "Variables to plot:" box, put the range of cells corresponding to the columns that contain the variables which you want to include in the trellis plot.
5. Make sure the choice for "Titles in the first row" or "No titles in the first row" is appropriate for the ranges you have entered at Steps 3 and 4.
6. Click "OK".

The trellis plot should now be created and shown in a new workbook in Excel.

2.10 STAR PLOTS

Star plots are so called because of the way in which they are meant to look like stars. A picture is created for each observation in the dataset, consisting of a point with rays coming out of it (supposedly like light radiating from a star). Figure 2.5 shows examples for three imaginary cases. Each star has five rays coming from it, each representing a different variable. For instance, in terms of the dataset described in Chapter 1, the ray at roughly the 11 o'clock position might represent height, the ray at 1 o'clock might represent weight, the ray at 4 o'clock might represent systolic blood pressure, the ray at 7 o'clock might represent diastolic blood pressure and the ray at just after 9 o'clock might represent pulse rate. The lengths of the rays represent the values recorded for each case for these five variables. Sometimes the ends of the rays are joined, as in Figure 2.6.

Whether the star plot is formed as in Figure 2.5 or Figure 2.6, the idea is that once formed, you can look at them and be able to see what characteristics are similar or different across a range of cases. You may look at the figures yourselves and wonder just how that might be done, and I have sympathy with your confusion. It is quite difficult to spot similarities and differences, but the key thing is that you have a greater chance of spotting such characteristics than if you are simply looking at a load of numbers.

FIGURE 2.5   Star plots for three imaginary cases.

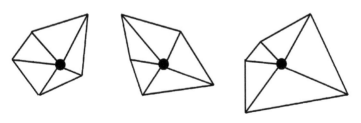

FIGURE 2.6   Joined star plots for three imaginary cases.

FIGURER 2.7   Chernoff faces for three imaginary cases.

2.11 CHERNOFF FACES

If you think star plots are a little odd, then you will be even more surprised by Chernoff faces. Here, rather than having rays coming from a star, each variable is represented by features on a face, as in Figure 2.7. Here, one variable might be represented by the width of the face, one might be represented the size of the face, one might be represented by the "smileyness" of the mouth, one might be represented by the length of the mouth and one might be represented by the distance between the eyes.

The idea, as with the star plots, is that you can look at the faces and spot similarities and dissimilarities. In theory, this should be easier than looking at star plots because we are used to looking at human faces. However, this does have a drawback because certain features (e.g. width of face) are more noticeable than others, and this attaches more importance to the variables being represented by these more noticeable features.

2.12 ANDREWS' PLOTS

Rather than produce a separate plot for each case in a dataset (as is done by star plots and Chernoff faces), Andrews' plots produce one graph on which is plotted a single curve for each case in the dataset. This curve is a summary of however many variables are being investigated, and is produced by applying the following formula:

$$x(t) = \frac{x_1}{\sqrt{2}} + x_2 \sin(t) + x_3 \cos(t) + x_4 \sin(2t) + x_5 \cos(2t)$$

$$+ x_6 \sin(3t) + x_7 \cos(3t) + \cdots$$

Now, to those readers not terribly at ease with calculus, the appearance of sine and cosine functions at this early stage of the book may be alarming. However, do not panic. We can be grateful to D. F. Andrews (1972) and a paper he published for coming up with this formula and showing that it can be used for the purpose explained in this section. You may also be glad to know that the Microsoft Excel add-in included with this book will create these plots for you, without you needing to do the calculus yourself.

In the formula, the x_1, x_2, x_3, etc. are the different variables being considered. If there are more than seven variables, the function continues in the same pattern as shown, with even-numbered variables associated with the sine function and odd-numbered variables with the cosine function. The "t" is what makes up the horizontal axis of the graph and ranges from $-\pi$ to $+\pi$. The reason for these starting and finishing values for t will be obvious to those familiar with sine and cosine functions. For those of you who are not, let me just say that this range covers all relevant values because once we get larger than $+\pi$, the sine and cosine functions start repeating what they give for values of t above $-\pi$. The curves would then be repeating themselves, which is rather pointless.

What we do in practice for each case in the dataset is calculate $x(t)$ for a whole range of values of t from $-\pi$ to $+\pi$, using the values of x_1, x_2, x_3, etc. that exist for that case. We then plot these $x(t)$ on a graph against t, as in Figure 2.8.

There is a further complication in that in the formula to create the curve, the first variable, x_1, has more influence over the appearance of the curve than

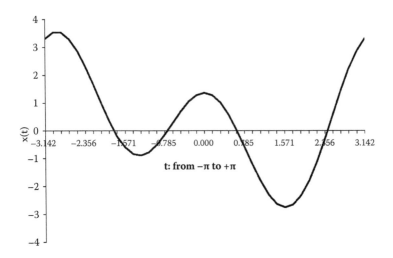

FIGURE 2.8 Andrews' plot for one case.

any of the other variables. Similarly, x_2 has more influence than any of x_3, x_4, etc., and x_3 has more influence of any of x_4, x_5, etc. Now, for datasets where it is the case that some variables are naturally more important than others, this is not a problem. When applying the formula, the most important variable must be assigned to be x_1, the next more important x_2 and so on.

However, what if there is no way to say that some variables are more important than others in a dataset? Fortunately, another multivariate statistical technique can come rushing to our aid. This is called *principal components analysis*, and we will be dealing with it in more detail in Chapter 4 when we discuss factor analysis. Those readers who wish to do so can of course dash off to that chapter now. However, if you are staying with me here for now, let me just briefly explain what principal components analysis does. Basically, principal components analysis takes the variables that are being used in the analysis, and creates a completely new set of variables by performing calculations based on the original variables. So, if originally there were three variables, for example, we end up with three new variables. Each of these new variables (say, z_1, z_2, z_3) is a simple linear combination of the original three variables: for instance, $z_1 = 0.39x_1 + 0.47x_2 + 0.79x_3$. The variable z_2 would also be a linear combination of the original variables but the multipliers for x_1, x_2 and x_3 would be different, and would be chosen so that the new variables z_1 and z_2 were not correlated with each other. Then z_3 would be another linear combination of x_1, x_2 and x_3 with multipliers chosen so that z_3 was not correlated with either z_1 or z_2.

Without going into the mathematics that prove it, once we have our three new variables z_1, z_2, and z_3 that are all independent of each other, then all the information that was originally contained in x_1, x_2 and x_3 would now be contained in z_1, z_2, z_3. Also (and this is the important bit for our Andrews' plots), z_1 would contain more of the information originally available from all of x_1, x_2 and x_3 than z_2 or z_3 and would therefore be the most important of the new variables. Similarly, z_2 would contain more information than z_3 and therefore be more important than z_3. We can then use z_1, z_2 and z_3 in our formula to create the Andrews' plots.

So now that we know how to create the curve in Figure 2.8, what does it tell us? Well, on its own, the answer is, "not a lot"! The power of Andrews' plots really comes when lots of cases are plotted. From the dataset discussed in Chapter 1, Figure 2.9 shows Andrews' plots for four variables – history of science score, politics score, sport score and history score – with principal components being used as there is no sensible way of saying that these scores can be put into an order of importance.

I will forgive you for wondering what that jumble of lines in Figure 2.9 is meant to represent. It does not look too illuminating. There are 100 curves in Figure 2.9, corresponding to the 100 cases in the dataset. However, closer inspection reveals that although most curves are jumbled up with each other, there are some which are not. These are unusual in some way. If they were not unusual, then the curves would be like the others. We can identify which case these unusual curves belong to and investigate further. In Figure 2.9, we see the curve created by case 17 highlighted (this has been done by simply hovering the mouse over the curve). Looking into the details of this case reveals an unusual pattern. For most cases in the dataset, if they score well in history of science, they also score well in history, and vice versa. However, for case 17, we find that he or she has the 16th highest score out of 100 for history of science but only the 98th highest score out of 100 for history.

Other unusual cases revealed by the Andrews' plots can also be investigated. The key thing to remember is that cases are identified as unusual because of an *unusual profile* across all the variables being considered, rather than just because they have unusually high or low scores on individual variables. It is this consideration of the profile which makes Andrews' plots a useful tool for multivariate data.

In Figure 2.9 it is only unusual cases which can be detected. In other circumstances, it may be the case that different groups of similar curves can be identified. This is then showing that the dataset can be divided into groups according to the variables under investigation. If you see this pattern in your Andrews' plots, you may want to further examine this aspect of your data using cluster analysis (see Chapter 5).

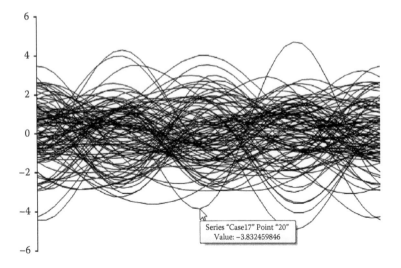

FIGURE 2.9   Andrews' plot for all cases.

2.13 A STEP-BY-STEP GUIDE TO OBTAINING ANDREWS' PLOTS USING THE EXCEL ADD-IN

1. You must have a column in Excel that contains the names by which your cases are known. These are called the "case identifiers". They may be names or codes that you can use to identify the different cases, or may be simply case numbers (e.g. case 1, case 2, etc.). You must also have columns of data containing the variables which you want to include in the construction of the Andrews' plots.
2. Go through the multivariate analysis add-in's menus until you get the dialogue box for Andrews' plots.
3. In the "Case identifiers:" (e.g. names) box, put the range of cells corresponding to the column in which the case names, labels or whatever (see Step 1) are located.
4. In the "Variables to plot:" box, put the range of cells corresponding to the columns that contain the variables to be used in the plots.
5. If there is no particular order of importance for the variables you wish to plot, then make sure the "Make Andrews' plots of principal components" option is selected. If the data do have an order of importance, then make sure the "Make Andrews' plots of data in order given" option is selected, but also make sure that in Excel, the first column from the left contains the most important variable, followed by the next most important, and so on.
6. Make sure the choice for "Titles in the first row" or "No titles in the first row" is appropriate for the ranges you have entered at Steps 3 and 4.
7. Decide whether you want colour or black-and-white plots, and select the appropriate option.
8. Click "OK".

The analysis should now take place. The results will be shown in a new workbook in Excel.

2.14 PRINCIPAL COMPONENTS PLOT

I am a bit reluctant to include this section in the book because it deals with producing a plot which deliberately throws away some of the information in the dataset. Put like this, it does sound like a daft idea but when you understand that what is retained in the plot is the most that can be contained in a two-dimensional plot and what is thrown away is as little as possible, then it does not sound so bad.

A principal components plot is just a two-dimensional scatterplot but rather than plotting two variables, it is the first two principal components that are plotted. What are "principal components"? Well, if you have read Section 2.12 about Andrews' plots, then you will have seen them mentioned there along with a brief explanation. For obvious reasons, I am not going to repeat myself in this section so if you have skipped Section 2.12, then do please go back and find my explanation there. There is one important difference between what we do for Andrews' plots and what we do for a principal components plot. With an Andrews' plot we transform the existing variables into principal components in such a way that if we have, say, four original variables, then we get four principal components and use them all when creating the Andrews' plot. For a principal components plot, we do the same thing, in that the number of principal components we create is the same as the number of variables we start with but we only use the first two principal components to create the plot. The remaining components, no matter how many of them there are, are simply discarded. This does not sound very satisfactory and indeed is not ideal. However, the principal components are created such that the first component contains more information about the data than any of the other components. The second component contains more information than any of the others apart from the first component. So, although we are throwing away some components, we are keeping the two most important ones.

You may already have reached the stage of thinking that, 'If we're throwing away information, is there any way we can have an idea of how much we are getting rid of?' The answer is, 'Yes'. Each principal component that is created has associated with it a thing called an *eigenvalue*. This is just a number resulting from the mathematics behind creating the principal components but it is directly related to the amount of information contained in its associated principal component. Let me explain further by pointing you towards an example. From the dataset discussed in Chapter 1, Figure 2.10 shows the principal components plot for four variables: history of science score, politics score, sport score and history score. But before we discuss that, let us look at Table 2.1. This shows us that the first component contains almost 62% of the information that is held by the four variables and the second component contains an additional 17%. Together, they contain almost 79% of the information in the original four variables. Thus, although we are throwing away information in the third and fourth components, we are only throwing away 21% of the overall information. This is not ideal, of course. It would be more comforting to find that we were throwing away a very small percentage but on the other hand, it could be a lot more that we were losing.

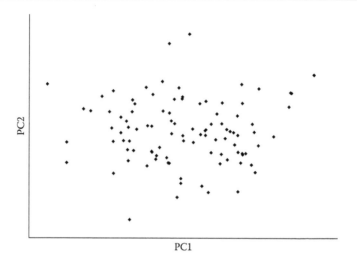

FIGURE 2.10   Principal components plot.

TABLE 2.1   Percentage of Information Accounted for by Components

COMPONENT	EIGENVALUE	PERCENTAGE OF INFORMATION IN ORIGINAL FOUR VARIABLES ACCOUNTED FOR BY COMPONENT	CUMULATIVE PERCENTAGE
1	2.470	61.753	61.753
2	0.685	17.123	78.876
3	0.509	12.722	91.598
4	0.336	8.402	100.000

Having dealt with the issue of how much information we are throwing away and thus how much is displayed in Figure 2.10, let us turn to what it is showing us. The first principal component (which in this case contains 62% of the information) is on the horizontal axis, and the second principal component (which in this case contains 17% of the information) is on the vertical axis. Cases in the dataset which are similar to each other (on the basis of these two components) will have points on the plot which are close to each other. Cases which are unusual in some way (on the basis of these two components) will have points on the plot which are not very near other points. As such, the principal components plot is able to show similar patterns to the Andrews' plots.

2.15 A STEP-BY-STEP GUIDE TO OBTAINING A PRINCIPAL COMPONENTS PLOT USING THE EXCEL ADD-IN

1. You must have a column in Excel that contains the names by which your cases are known. These are called the "case identifiers". They may be names or codes that you can use to identify the different cases, or may be simply case numbers (e.g. case 1, case 2, etc.). You must also have columns of data containing the variables which you want to include in the construction of the principal components plot.
2. Go through the multivariate analysis add-in's menus until you get the dialogue box for principal components plots.
3. In the "Case identifiers": (e.g. names) box, put the range of cells corresponding to the column in which the case names, labels or whatever (see Step 1) are located.
4. In the "Variables to plot": box, put the range of cells corresponding to the columns that contain the variables to be used in the plots.
5. Make sure the choice for "Titles in the first row" or "No titles in the first row" is appropriate for the ranges you have entered at Steps 3 and 4.
6. Click "OK".

2.16 MORE INFORMATION

Over the years, there have been many different graphical displays developed for multivariate data. Those shown in this chapter are some of the better-known ones. Readers interested in finding out more about this topic may like to look at Manly (2005) and Brown et al. (2012).

Multivariate Tests of Significance

3

3.1 WHY DO I WANT TO DO MULTIVARIATE TESTS OF SIGNIFICANCE?

If you have reached this stage of reading this book, then it is likely that you already know something about what can be called *univariate* tests of significance. These are just the *t*-tests, *F*-tests, and simple ANOVA and related methods of analysis. Here we are calling these univariate because they only have one variable which is being analysed at a time. For instance, we may want to know if the blood pressure measurements for two groups of patients are the same or different. Although we have two bits of information on each patient (blood pressure reading and to which group they belong), it is only one variable – the blood pressure – which has its data subjected to various calculations. Further information about univariate tests is beyond the scope of this book, and if you want to know more, I can recommend Field (2009). However, there are an immense number of other books available that discuss these subjects.

Having read the above, you have probably already guessed that multivariate tests of significance are tests which have the same aims as univariate tests but have more than one variable being analysed. Hence, expanding the example above, rather than just seeing whether or not two groups of patients differ in respect to blood pressure readings, a multivariate test could see whether or not they differ in respect to both blood pressure and pulse rate. You may now be wondering why for this example we do not simply do two univariate tests: one for blood pressure and one for pulse rate. This is straying into territory that is discussed more thoroughly in Section 3.4, but for now let me simply explain that a multivariate test examines both of these variables at the same time, taking into account any relationship between them, whereas doing two univariate tests ignores any relationship that exists. On the basis that deliberately ignoring information concerning our variables cannot be a good idea, we would prefer the single multivariate test over the two univariate tests.

3.2 WHAT DATA DO I NEED FOR MULTIVARIATE TESTS OF SIGNIFICANCE?

As with univariate tests, we need continuous data to be able to undertake multivariate tests of significance, or at least data which can be regarded as continuous. A discussion of what continuous data is can be found in Chapter 1.

However, having said that all we need is data which are continuous or can be treated as continuous, I must point you towards the assumptions that are necessary for the multivariate tests of significance that are detailed in the various subsequent sections of this chapter. If the data do not satisfy these assumptions, then it is irrelevant whether or not the data are continuous or can be treated as continuous.

As well as continuous data, we also need to have information about groups which exist in the data. The multivariate tests of significance outlined in this chapter are all about comparing groups with each other. We must therefore have some means of identifying which cases are in which groups.

3.3 THE REST OF THIS CHAPTER

There are many different multivariate tests of significance, and I am not attempting to give a comprehensive list here. Instead, I am showing just a few tests that between them do the job of testing hypotheses concerning means and covariance matrices. The reason I have included these rather than other tests which aim to do the same thing is because of the clear parallels they have with univariate analyses. This does not necessarily make them the "best" tests but they do the job in an understandable way and simply "getting the job done" is often sufficient.

The sections that follow in this chapter deal with this comparing of means and variation. In the univariate case, we have different tests that we use, depending on whether we are comparing two means or variances or we are comparing more than two means or variances. It is the same in the multivariate

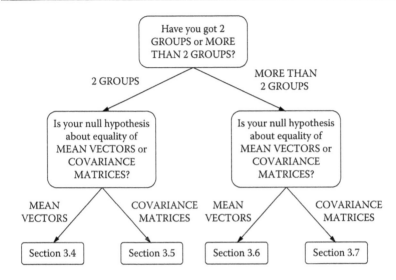

FIGURE 3.1   Flowchart for test.

framework. Hence, we start by comparing means for two groups in Section 3.4 and comparing variation for two groups in Section 3.5. We then follow on to compare means for more than two groups in Section 3.6 and variation for more than two groups in Section 3.7. As this all sounds a bit complex, I have included a flowchart in Figure 3.1 to help. The chapter finishes by pointing you towards more information on the topic, should you want it, in Section 3.8.

3.4 COMPARING TWO VECTORS OF MEANS

3.4.1 What Are We Testing?

From the dataset discussed in Chapter 1, let us concentrate on three variables: systolic blood pressure, diastolic blood pressure and pulse rate. Let us also consider the two groups in the dataset defined by gender: male and female. The hypotheses that we wish to examine are as follows.

- H_0: males and females have the same means for systolic blood pressure, diastolic blood pressure and pulse rate.
- H_1: H_0 is not true.

3.4.2 What Is a Vector of Means?

Do not be scared by the word *vector*. It is indeed a mathematical term, but essentially it is just another word for "list". So, when we say we are going to compare two vectors of means, it simply implies that we have two groups and we want to see if the list of means for one group has the same values as the list of means for the other group.

3.4.3 Univariate Tests

In a univariate setting, the vector or list of means for each group would simply contain one value: the mean for whatever variable is being analysed. We would then carry out a t-test (with pooled variance or with separate variances if the assumption of equal variances was not appropriate). If the data were not normally distributed, we would consider using a non-parametric method of analysis instead.

So what would happen if we carried out three separate t-tests for the systolic blood pressure, diastolic blood pressure and pulse rate? With H_0: males and females have the same means, H_1: males and females have different means, we are operating with a two-sided test. Histograms of the variables indicate that they follow a normal distribution sufficiently enough, and as the standard deviations are sufficiently similar, pooled variance t-tests are appropriate. The results of the three analyses are shown in Table 3.1 which shows the relevant means, standard deviations, t-statistics and p-values. We conclude that we have insufficient evidence to reject the null hypothesis for systolic blood pressure and pulse rate at the 5% level of significance (p-values are larger than 0.05), but we do have sufficient evidence to reject the null

TABLE 3.1 t-Tests of Results

	MALE		FEMALE			
VARIABLE	MEAN	STD. DEV.	MEAN	STD. DEV.	t-STATISTIC	p-VALUE
Systolic blood pressure	112.25	9.782	110.54	10.468	0.843	0.401
Diastolic blood pressure	64.88	6.313	62.37	5.573	2.111	0.037
Pulse rate	70.94	9.336	68.85	9.566	1.105	0.272

hypothesis for diastolic blood pressure at the 5% level of significance (*p*-value is less than 0.05).

Now, one could argue that we have a solution to the hypotheses set out in Section 3.4.1 without the need to conduct a multivariate test of significance: we can reject the null hypothesis because males and females have different diastolic blood pressures. However, by conducting three tests each at the 5% level of significance, we have an inflated chance of making what is sometimes called a type I error: rejecting the null hypothesis when in reality for the population (as opposed to the sample for which we have data available for analysis), the null hypothesis is true. When we rejected the null hypothesis for the diastolic blood pressure above, we did so on the basis that the *p*-value was 0.037 (3.7%), less than 0.05 (5%). What this 3.7% *p*-value is telling us is that there is a 3.7% chance that what we are observing for the diastolic blood pressure could be observed by chance if in fact the true means for males and females in the population were the same. When drawing a conclusion about significance, we are in effect saying that because 3.7% is so small (that is, less than the 5% significance level cut-off we are using), we do not believe the differences observed could have happened by chance, and must be due to males and females having different mean values in the population as a whole.

By conducting three tests, each at the 5% level of significance, we are allowing ourselves to have a 5% risk of making a type I error for each test. Overall, if we assume that the three tests are independent of each other (not a valid assumption, but please wait until a later paragraph for me to pick up this issue again), this means that we have a 14.26% chance of making a type I error at some point in our triple analysis ($1 - (1 - 0.05)^3 = 0.1426$). This does not sound as acceptable as a 5% chance that we normally deal with in hypothesis testing!

One solution to this inflated chance of making a type I error would be to reduce the significance level we use for each individual *t*-test to 1.695%, meaning that the overall chance of making a type I error is 5% ($1 - (1 - 0.01695)^3 = 0.05$). Doing this would now mean that we would accept the null hypothesis for the diastolic blood pressure and conclude that instead of rejecting the null hypothesis of Section 3.4.1, we now accept it!

Now, above I promised to return to the issue of whether the tests can be considered independent of each other. The calculations of the 14.26% and the 1.695% in the preceding paragraphs were based on this assumption. But how realistic is this? If we examine the data, we find that the correlation between systolic and diastolic blood pressure in our sample is 0.59. Thus, if there are really differences between males and females for diastolic blood pressure, then surely that might imply that there could be a difference for systolic blood pressure as well. Looking again at Table 3.1, we see that the *p*-value for the systolic blood pressure *t*-test is 40.1%. Although this is not at all near the 5% level of significance cut-off, it is nevertheless not terribly high. And indeed this is what we would expect: where two variables have a reasonable correlation (such as systolic and diastolic blood pressure here), we would expect the *p*-values resulting from identical tests on them to be similar.

So, if we accept that the assumption of independence of the *t*-tests shown in Table 3.1 is not sustainable, what is the probability of at least one type I error previously calculated to be 14.26%? What should we use instead of the 1.695% level of significance above to try and adjust the overall probability of a type I error to 5%? To all intents and purposes, the answer to both of these questions is, 'We do not know'. It is possible to do rather gruesome calculations to give some sort of idea but this is not a terribly practical way of proceeding.

Instead, a neater solution would be to conduct a single multivariate test that assesses the equality of the means in one go. We could then use a 5% level of significance for this one test. This approach would also have an additional advantage because we could take into account information about the relationships between the three variables being tested. When we did the three univariate tests above, we were ignoring the fact that we had access to this information, and that cannot be a good thing! In Section 3.4.5 we discuss the most popular way of conducting a multivariate test of significance to compare two vectors or means: Hotelling's T^2 test. But before that, in Section 3.4.4 we discuss the important topic of what assumptions are being made when conducting the test.

3.4.4 Assumptions Made for Multivariate Test

Before undertaking any hypothesis test, it is wise to consider the assumptions being made. For Hotelling's T^2 test, there are three key assumptions, as follows.

1. The cases in the data are independent of each other.
   - In an ideal world, the cases that make up your data would be a random selection from the population of interest in your study. In the real world, your sample is likely to consist of those who responded to a survey, or those who you were able to interview or cases that for some other reason were available for data collection purposes. It is important that this data collection stage is undertaken with an appreciation that for most statistical analyses, you will need to assume that the data are independent of each other. You should thus try and plan your

data collection so that your final sample contains cases that are as independent of each other as is practically possible. So, for example, if you are collecting data from one particular person, try to avoid also collecting data from their partner or friend, even if it is convenient to do so at the same time. For the data we are using here, we can be happy that the independence assumption is valid.

2. The data come from a multivariate normal distribution.
   - In a univariate setting, an assumption that often must be made is that a particular variable has a normal distribution. This can be checked by looking at a histogram of the variable and seeing if it looks roughly like a normal distribution. For a multivariate test, it is difficult to create a three-dimensional distribution for two variables and see if it looks like a normal distribution from all relevant angles. It is even more difficult to create a four-dimensional distribution for three variables or deal with more dimensions for more variables! In practice, if each of the variables being investigated in the multivariate test has a normal distribution itself, then we can be reassured that jointly they will have a multivariate normal distribution. Histograms of all three variables being used here reveal patterns near enough to a classic normal distribution shape for us to be happy with this assumption.

3. The covariance matrices for the two populations being investigated are the same.
   - When conducting a univariate t-test using pooled variances, one must assume that the variance is the same in both populations of interest. For Hotelling's T^2 test, the same assumption must be made, but now instead of just considering variances, we must consider the entire covariance matrix. The subject of what a covariance matrix consists of is addressed in Section 3.4.5. For now, let me state that what we need to assume is not only that the variance for each variable is the same in both populations, but also the various covariances. The validity of this assumption can be checked by simple inspection of the variances and covariances obtained from the data available for analysis. We can see from this that the standard deviations (and thus variances) are fairly similar for males and females in our sample for all three variables. Inspection of the covariances between the variables reveals figures that are sufficiently similar for both genders for us to be content with this assumption.

MATRIX 3.1   Example covariance matrix

$$
\begin{bmatrix}
\begin{array}{lll}
\text{variance of} & \text{covariance for} & \text{covariance for} \\
\text{variable 1} & \text{variables 1 and 2} & \text{variables 1 and 3} \\
\text{covariance for} & \text{variance of} & \text{covariance for} \\
\text{variables 1 and 2} & \text{variable 2} & \text{variables 2 and 3} \\
\text{covariance for} & \text{covariance for} & \text{variance of} \\
\text{variables 1 and 3} & \text{variables 2 and 3} & \text{variable 3}
\end{array}
\end{bmatrix}
$$

3.4.5 What Is a Covariance Matrix?

The word *matrix* sounds like another scary mathematical term but it really just means a table with rows and columns. An example of a covariance matrix is given in Matrix 3.1 for a three-variable situation. It is simply a grid showing the variances for each of the variables down the diagonal from top left to bottom right and covariances in the off-diagonal positions. The first column and first row is where you will find the covariances involving the first variable. The second variable is associated with entries in the second row and column, and the third variable corresponds to entries in the third row and column. It is easy to see what the pattern will be for situations where there are more than three variables.

The matrix is also symmetric. That is, an entry in the first row, second column will be identical to the entry in the second row, first column. This is because the covariance between variable 1 and variable 2 is the same as the covariance between variable 2 and variable 1. The symmetric nature of the covariance matrix can be seen more clearly in Section 3.4.6 where the covariance matrices for the three variables being examined here are given.

The practice of putting brackets around the covariance matrix is one of convention and need not be of concern. It is a convenient way of showing that the matrix is a single entity. Without the brackets, it might not be clear that all the figures belong together.

3.4.6 Hotelling's T^2 Test

We start with the technical bit so that you know what is going on when you do Hotelling's T^2 test. It is not essential that you feel comfortable in undertaking each and every step of the calculations yourself, as the add-in provided with this book will do it for you. However, it is a good idea if you have at least some idea what is going on behind the scenes.

The initial stage is to combine the covariance matrices for the two groups in our example: males and females. The separate covariance matrices are below. The first row/column corresponds to the systolic blood pressure, the second row/column refers to the diastolic blood pressure and the third row/column refers to the pulse rate. The **S** notation is to indicate that these are covariance matrices obtained from our samples of data.

$$
S_{male} = \begin{bmatrix} 95.681 & 37.266 & -4.303 \\ 37.266 & 39.856 & 28.439 \\ -4.303 & 28.439 & 87.166 \end{bmatrix}, \quad S_{female} = \begin{bmatrix} 109.587 & 33.701 & -20.210 \\ 33.701 & 31.060 & 14.979 \\ -20.210 & 14.979 & 91.505 \end{bmatrix}
$$

In a univariate setting, when we want to combine two variances to undertake a pooled variances t-test, we calculate a weighted average:

$$
s^2 = \frac{(n_1 - 1)s_1^2 + (n_2 - 1)s_2^2}{n_1 + n_2 - 2}
$$

where s_1^2 is the sample variance from the first group; s_2^2 is the sample variance from the second group; and n_1, n_2 are the number of cases in the data for the groups.

To pool the covariance matrices for Hotelling's T^2 test, we follow the same approach:

$$
S = \frac{(n_{male} - 1)S_{male} + (n_{female} - 1)S_{female}}{n_{male} + n_{female} - 2}
$$

To multiply a matrix by a constant, as is required here, all you do is multiply each individual element of the matrix (that is, each number in the matrix) by the constant. With n_{male} being 48 and n_{female} being 52 for this data, we obtain the following pooled covariance matrix:

$$
S = \begin{bmatrix} 102.918 & 35.411 & -12.581 \\ 35.411 & 35.279 & 21.434 \\ -12.581 & 21.434 & 89.424 \end{bmatrix}
$$

To calculate Hotelling's T^2 statistic, we use the formula

$$
T^2 = \frac{n_{male} n_{female} \left(\bar{x}_{male} - \bar{x}_{female} \right)^T S^{-1} \left(\bar{x}_{male} - \bar{x}_{female} \right)}{n_{male} + n_{female}}
$$

where $\bar{x}_{male}$ is the mean vector for the first group, $\bar{x}_{female}$ is the mean vector for the second group and S^{-1} is the inverse of the pooled covariance matrix. The computation of this from S is not straightforward and is not discussed here.

However, it is a standard mathematical technique, and further information about how we obtain S^{-1} from S can be found in many places. Here,

$$\bar{x}_{male} = \begin{pmatrix} 112.25 \\ 64.88 \\ 70.94 \end{pmatrix} \quad and \quad \bar{x}_{female} = \begin{pmatrix} 110.54 \\ 62.37 \\ 68.85 \end{pmatrix}$$

$$so \; \bar{x}_{male} - \bar{x}_{female} = \begin{pmatrix} 1.71 \\ 2.51 \\ 2.09 \end{pmatrix}$$

by simply taking each element of $\bar{x}_{female}$ away from the corresponding element of $\bar{x}_{male}$. The first row corresponds to the systolic blood pressure, the second row refers to the diastolic blood pressure and the third row refers to the pulse rate.

The "T" superscript in the $\left(\bar{x}_{male} - \bar{x}_{female}\right)^T$ part of the formula for Hotelling's T^2 statistic means that the vector of differences is written as a row: $\left(\bar{x}_{male} - \bar{x}_{female}\right)^T = \begin{pmatrix} 1.71 & 2.51 & 2.09 \end{pmatrix}$. Carrying out the appropriate calculations gives a value of $T^2 = 4.708$.

Once Hotelling's T^2 has been calculated, we create an F-statistic so that we can compare it with an F-distribution. To create the F-statistic, we use the formula

$$F = \frac{(n_1 + n_2 - p - 1)T^2}{(n_1 + n_2 - 2)p}$$

where n_1 and n_2 are defined as above and p is the number of variables involved (3 in this example). In our example here, this gives $F = 1.537$.

The relevant numbers of degrees of freedom to use when we compare this F-statistic with an F-distribution are p and $n_1 + n_2 - p - 1$, which here are 3 and 96, respectively. This yields a p-value of 0.210.

As the p-value is more than 5%, we conclude that we have insufficient evidence to reject the null hypothesis of Section 3.4.1 at the 5% level of significance.

We have arrived at this conclusion whilst maintaining our 5% chance of a type I error and also taking into account the information we know about the relationships between the variables. Of course, if we had rejected the null hypothesis here, it would not mean that all the mean values are different for males and females – it could be that the mean values for some of the variables were the same and for others were different. In order to examine what differences did exist, we would have to investigate further, possibly using univariate *t*-tests.

3.4.7 A Step-by-Step Guide to Comparing Two Vectors of Means Using the Excel Add-In

1. You must have a column in Excel that contains the names by which your cases are known. These are called the "case identifiers". They may be names or codes that you can use to identify the different cases, or may be simply case numbers (e.g. case 1, case 2, etc.). You must also have columns of data in Excel containing the variables whose means you want to include in the comparison and a column that tells Excel which of the two groups each case is in.
2. Go through the multivariate analysis add-in's menus until you get the dialogue box for multivariate tests of significance.
3. In the "Case identifiers:" box, put the range of cells corresponding to the column in which the case names, labels or whatever (see Step 1) are located.
4. In the "Variables to use in analysis:" box, put the range of cells corresponding to the variable you are using in the analysis.
5. In the "Group identifiers:" box, put the range of cells corresponding to the column that indicates which of the two groups each case in the dataset is in.
6. Make sure the Yes/No choice for "Variable names in first line of data?" is appropriate for the ranges you have entered at Steps 3, 4 and 5.
7. Make sure "Means" is selected for "Compare vectors of means or covariance matrices?".
8. Click "OK".

The analysis should now take place. The results will be shown in a new workbook in Excel.

3.5 COMPARING TWO COVARIANCE MATRICES

3.5.1 What Are We Testing and How?

When we set about comparing two covariance matrices, we are of course talking about comparing the variances and covariances contained in the matrices. So if we decide that two covariance matrices are similar to each other, then what we are really saying is that the variances in the first matrix are similar to the corresponding variances in the second, and also that the covariances in the first matrix are similar to the corresponding covariances in the second matrix.

In the univariate world, the equivalent test is to compare two variances. We ask if the variation in a variable for one group is the same as it is for another group. This is a question that is asked when undertaking a two-sample t-test: if the variances for the two groups can be assumed to be equal, then the pooled-variance t-test can be undertaken. If that assumption is not justified, then a version of the test is used which keeps the variances separate. One way of assessing whether or not the assumption is reasonable is to perform Levene's test. This operates by looking at how far each value for a variable is from the mean or median of its group. If a group has a high variance, then the values in that group will be, on average, quite far away from its mean/median and so its average deviation will be high. If a group has a low variance, then the values in that group will be, on average, quite near its mean/median and so its average deviation will be low. Levene's test simply compares these averages using a standard t-test. If the groups have variances which are sufficiently different, then the t-test will give a p-value small enough to reject the null hypothesis of equal variances. Of course, when calculating the average deviation, something must be done to avoid negative deviations cancelling out positive deviations. For Levene's test, absolute values are taken (that is, negative deviations are made positive by having their minus signs removed).

In a multivariate setting, the idea behind Levene's test can again be used. Here, we again use the dataset discussed in Chapter 1 and the three variables: systolic blood pressure, diastolic blood pressure and pulse rate. We again consider the two groups in the dataset defined by gender: male and female. The hypotheses that we wish to examine are as follows.

- H_0: males and females have the same covariance matrices for systolic blood pressure, diastolic blood pressure and pulse rate.
- H_1: H_0 is not true.

3.5.2 Assumptions Made

Again, as we are undertaking a hypothesis test, it is wise to consider the assumptions being made. For the multivariate Levene's test, there are three key assumptions, as follows:

1. The cases in the data are independent of each other.
   - See Section 3.4.4 for a discussion of this assumption.
2. The data come from a multivariate normal distribution.
   - Again, please read the relevant part of Section 3.4.4 for a discussion of this assumption but here, instead of the original data for systolic blood pressure, diastolic blood pressure and pulse rate, it is now the absolute deviations from the mean or median which need to be normally distributed.
3. The covariance matrices for the two populations being investigated are the same.
   - It is perhaps a bit odd to have to make an assumption about two covariance matrices being the same when the object of the test is to compare two covariance matrices. However, here the assumption is about the covariance matrices of the absolute deviations rather than the original variables. As described in Section 3.4.4, an assessment of this assumption can best be made by looking at the two covariance matrices involved and seeing whether or not they look similar.

3.5.3 Multivariate Levene's Test

As with Hotelling's T^2 test, we start with the technical bit so that you know what is going on when the test is undertaken. The Excel add-in provided with this book will do the difficult work for you, but it is a good idea for you to know something of what is going on. In fact, so much is a repeat of a Hotelling's T^2 test that there is not a lot of additional technical detail to present here.

The multivariate Levene's test starts by taking the group means or medians away from each variable and taking absolute values. Thus, we take the mean or median systolic blood pressure for the males away from actual values of systolic blood pressure for each of the males, and ignore any minus signs. We also take away the mean or median systolic blood pressure for the females from each of the actual values of systolic blood pressure for the females and again ignore any minus signs. This process is then repeated for the diastolic blood pressure and pulse rate.

You have no doubt realised by now that I have been very non-committal about whether it is the mean or median that should be deducted. This is deliberate but I will now explain why. In Section 3.5.2 we stated that the absolute deviations from the mean or median had to follow a multivariate normal distribution. In reality, no variable is going to have a pattern which looks exactly like a normal distribution, but hopefully we will have distributions that are "not too bad" when compared with normality. If this is the case, then when choosing between the mean and median for the multivariate Levene's test, we might as well use the mean. A philosophical advantage it has over the median is that all the data for the group in question are used to calculate it, whereas the median is just one value that happens to be in the middle of the distribution. However, this advantage is also its disadvantage when it comes to data that have a distribution which is not as close to normal as one would wish. In these cases of dubious normality, the calculation of the mean might be influenced by some unusually small or large values. The median, however, is little affected by these unusual values, and is then to be preferred when undertaking the multivariate Levene's test. To use technical language, using the median instead of the mean leads to a more *robust* test. Because of this robustness property, we proceed here by taking away the median values.

The vectors of medians (as opposed to the vectors of means shown in Section 3.4.6) are

$$\tilde{x}_{male} = \begin{pmatrix} 112.0 \\ 66.0 \\ 70.0 \end{pmatrix} \quad \text{and} \quad \tilde{x}_{female} = \begin{pmatrix} 111.0 \\ 63.0 \\ 70.5 \end{pmatrix}$$

The first row corresponds to systolic blood pressure, the second row refers to diastolic blood pressure and the third row refers to pulse rate. Once these values are taken away from the respective variables for each group and absolute values taken, we have the following vectors of mean absolute differences and associated covariance matrices:

$$\overline{\text{abs. diff}}_{male} = \begin{pmatrix} 7.333 \\ 5.208 \\ 7.563 \end{pmatrix} \quad \overline{\text{abs. diff}}_{female} = \begin{pmatrix} 8.231 \\ 4.058 \\ 7.385 \end{pmatrix}$$

$$\mathbf{S}_{\text{abs. diff,male}} = \begin{bmatrix} 40.823 & 8.270 & -8.574 \\ 8.207 & 13.445 & 2.540 \\ -8.574 & 2.540 & 29.656 \end{bmatrix}$$

$$\mathbf{S}_{\text{abs. diff,female}} = \begin{bmatrix} 40.730 & 4.477 & -7.287 \\ 4.477 & 14.683 & 4.477 \\ -7.287 & 4.477 & 38.692 \end{bmatrix}$$

Inspection of the distributions of the absolute deviations reveals patterns similar enough to normal distributions to be happy with this necessary assumption, and the covariance matrices are also similar enough for us to be happy to proceed. We then proceed to undertake a Hotelling's T^2 test as in Section 3.4.6.

We obtain a T^2 statistic of 4.198 and thus an F-statistic of 1.371 with 3 and 96 degrees of freedom. This yields a p-value of 0.256.

As the p-value is more than 5%, we conclude that we have insufficient evidence to reject the null hypothesis of Section 3.5.1 at the 5% level of significance. We are thus content that the covariance matrices for males and females are sufficiently similar as to be regarded as the same. This gives added justification for the assumption that was made in Section 3.4 when we were comparing the two vectors of means for males and females.

3.5.4 A Step-by-Step Guide to Comparing Two Covariance Matrices Using the Excel Add-In

1. You must have a column in Excel which contains the names by which your cases are known. These are called the "case identifiers". They may be names or codes that you can use to identify the different cases, or may be simply case numbers (e.g. case 1, case 2, etc.). You must also have columns of data in Excel containing the variables whose covariance matrices you want to include in the

comparison and a column that tells Excel which of the two groups each case is in.

2. Go through the multivariate analysis add-in's menus until you get the dialogue box for multivariate tests of significance.
3. In the "Variables to use in analysis:" box, put the range of cells corresponding to the variable you are using in the analysis.
4. In the "Group identifiers:" box, put the range of cells corresponding to the column that indicates which of the two groups each case in the dataset is in.
5. Make sure the Yes/No choice for "Variable names in first line of data?" is appropriate for the ranges you have entered at Steps 3 and 4.
6. Make sure "Covariance matrices" is selected for "Compare vectors of means or covariance matrices?"
7. Click "OK".

The analysis should now take place. The results will be shown in a new workbook in Excel.

3.6 COMPARING MORE THAN TWO VECTORS OF MEANS

3.6.1 What Are We Testing, and How?

When we carried out the Hotelling's T^2 test in Section 3.4, we were comparing the mean vectors for two groups. When we want to compare the mean vectors for more than two groups, we have to use something different. It is a similar story in the univariate framework: to compare two means we use a t-test, but to compare more than two means, we use one-way analysis of variance.

The test described in this section does have similarities to the univariate one-way analysis of variance. In that method of analysis, we have to consider *sums of squares* and ratios of *mean squares*. In the Wilks' lambda test described in Section 3.6.3, we again consider matrices that are sums of squares.

We are again going to consider the three variables used in Sections 3.4 and 3.5 from the dataset described in Chapter 1, namely systolic blood pressure, diastolic blood pressure and pulse rate. However now, rather than consider two groups defined by gender, we consider the four groups defined by smoking history: "never smoked", "occasional smoker", "ex-smoker", "current smoker". The hypotheses that we wish to examine are as follows.

- H_0: groups defined by smoking history have the same means for systolic blood pressure, diastolic blood pressure and pulse rate.
- H_1: H_0 is not true.

3.6.2 Assumptions Made

The assumptions being made are identical to those required for Hotelling's T^2 test, as follows. See Section 3.4.4 for more details.

1. The cases in the data are independent of each other.
2. The data come from a multivariate normal distribution.
3. The covariance matrices for the different populations being investigated are the same.

3.6.3 Wilks' Lambda Test

To give you an idea of the logic behind the Wilks' lambda test, let us start by showing the formula for the test statistic. It may look weird and worrisome but do not be overly concerned. The logic is relatively straightforward, and the Excel add-in provided with this book will do all the necessary calculations.

$$\Lambda = |W| / |T|$$

In order to assess the size of this test statistic, a further adjustment must be made, as below. The resulting φ can then be compared with a chi-square distribution with $p(m-1)$ degrees of freedom.

$$\varphi = \left[n - 1 - \tfrac{1}{2}(p - m) \right] \ln[\Lambda]$$

In this formula, n is the total number of cases in the dataset (100 in this case), p is the number of variables being dealt with (3 in this case), and m is the number of groups (4 in this case). The **T** can be called the *total sum of squares and cross-products matrix;* but rather than let you worry about what this means, let me tell you it is just the covariance matrix (the sort of thing described in Section 3.4.5), but without any dividing being done. That is, the formula for a sample variance in the covariance matrix is

$$\frac{1}{n-1}\sum_{i=1}^{n}(x_i - \bar{x})^2$$

and for a sample covariance is

$$\frac{1}{n-1}\sum_{i=1}^{n}(x_{1i} - \bar{x}_1)(x_{2i} - \bar{x}_2)$$

In creating the **T** matrix, these variances and covariances are calculated but the division by $n - 1$ does not take place. When doing this calculating of **T**, the entire dataset is used in one go. That is, the means $\bar{x}_1$, $\bar{x}_2$, etc. are calculated for all the data together (ignoring the fact that they are in groups), and the variances and covariances are created using all the cases in the dataset and deliberately not dividing by $n - 1$.

The **W** matrix is the *within-samples sum of squares and cross-products matrix*. That is, the covariance matrix without the divisors (i.e. the **T** matrix) is created separately for each group (the four types of smoking habit here). The means $\bar{x}_1$, $\bar{x}_2$, etc. are now calculated separately for each group, and the individual group-specific covariance matrices (without dividing by $n - 1$) are calculated using them. These individual matrices are then simply added together to get the final **W** matrix.

The notation $|\mathbf{T}|$ and $|\mathbf{W}|$ is used to indicate the determinant of the **T** and **W** matrices, respectively. It is beyond the scope of this book to describe the determinant in detail – it is a standard mathematical technique which you can find out about in many different places if you wish. For the present purposes, it is sufficient to explain that the determinant is a single-number measure that summarises the matrix. Of particular relevance is the fact that if **T** and **W** are exactly the same, then $|\mathbf{T}|$ and $|\mathbf{W}|$ will be identical, meaning that $\ln\left[|\mathbf{W}|/|\mathbf{T}|\right] = \ln[1] = 0$ and so $\varphi = 0$.

But under what circumstances would **T** and **W** be the same? The answer to this is, "If the null hypothesis of Section 3.6.1 is true". Each group contributes to the **W** matrix via calculations that use the vector of means of the variables which are specific to each group. If these vectors of means are identical across the groups, then all the $\mathbf{W}$ calculations are using the same vector of means, and this will be identical to the vector of means used to calculate the $\mathbf{T}$ matrix. The $\mathbf{T}$ and $\mathbf{W}$ matrices will then be identical.

Of course, in reality, even if the null hypothesis of Section 3.6.1 is true, the data in our sample are never likely to give a vector of means that is absolutely identical in each group. However, the more similar the vectors of means are, the more similar $\mathbf{T}$ and $\mathbf{W}$ will be, and hence the nearer to zero $\ln\left[|\mathbf{W}|/|\mathbf{T}|\right]$ will be. The more dissimilar the vector of means, the more dissimilar $\mathbf{T}$ and $\mathbf{W}$ will be, and the further from zero $\ln\left[|\mathbf{W}|/|\mathbf{T}|\right]$ will be. Multiplying $\ln\left[|\mathbf{W}|/|\mathbf{T}|\right]$ by $\left[n-1-\frac{1}{2}(p-m)\right]$ gives a test statistic φ which can be compared with a chi-square distribution with $p(m-1)$ degrees of freedom.

Here, the four vectors of means for the smoking habit groups are

$$\bar{x}_{\text{never smoked}} = \begin{pmatrix} 107.15 \\ 59.55 \\ 65.30 \end{pmatrix} \quad \bar{x}_{\text{occasional smoker}} = \begin{pmatrix} 109.75 \\ 63.44 \\ 73.81 \end{pmatrix}$$

$$\bar{x}_{\text{ex-smoker}} = \begin{pmatrix} 111.70 \\ 64.75 \\ 71.20 \end{pmatrix} \quad \bar{x}_{\text{current smoker}} = \begin{pmatrix} 119.17 \\ 69.38 \\ 73.67 \end{pmatrix}$$

Inspection of the distribution of the variables being used shows that they follow a normal distribution sufficiently for us to be happy with the assumption of multivariate Normality. We also inspect the covariance matrices to see if they are sufficiently similar for us to be satisfied with the equal covariance matrices assumption, and conclude that although there are some differences, they are not so great that they mean we cannot undertake the test:

$$\mathbf{S}_{\text{never smoked}} = \begin{bmatrix} 112.438 & 31.044 & -39.482 \\ 31.044 & 27.126 & 9.856 \\ -39.482 & 9.856 & 97.241 \end{bmatrix}$$

$$\mathbf{S}_{\text{occasional smoker}} = \begin{bmatrix} 51.267 & 7.583 & -15.517 \\ 7.583 & 21.729 & 8.154 \\ -15.517 & 8.154 & 56.429 \end{bmatrix}$$

$$\mathbf{S}_{\text{ex-smoker}} = \begin{bmatrix} 61.589 & 6.763 & -26.463 \\ 6.763 & 15.250 & 15.263 \\ -26.463 & 15.263 & 71.958 \end{bmatrix}$$

$$\mathbf{S}_{\text{current smoker}} = \begin{bmatrix} 70.406 & 15.152 & -11.159 \\ 15.152 & 19.897 & 6.435 \\ -11.159 & 6.435 & 60.928 \end{bmatrix}.$$

Calculation of the $\mathbf{W}$ and $\mathbf{T}$ matrices gives

$$\mathbf{W} = \begin{bmatrix} 7943.633 & 1801.450 & -2532.017 \\ 1801.450 & 2131.213 & 944.713 \\ -2532.017 & 944.713 & 7407.371 \end{bmatrix}$$

and

$$\mathbf{T} = \begin{bmatrix} 10159.040 & 3577.480 & -1143.600 \\ 3577.480 & 3614.510 & 2231.550 \\ -1143.600 & 2231.550 & 8872.750 \end{bmatrix}.$$

The (rather large!) determinants that result from these are $|\mathbf{W}| = 71{,}993{,}825{,}597$ and $|\mathbf{T}| = 138{,}673{,}536{,}666$, giving $\Lambda = 0.519$. Put together with the rest of the formula for φ gives $\varphi = 65.266$ with $p(m-1) = 9$ degrees of freedom. This yields a p-value of 0.000 to three decimal places.

As the p-value is less than 5%, we conclude that we have sufficient evidence to reject the null hypothesis of Section 3.6.1 at the 5% level of significance. We thus say that the four smoking habit groups do differ in their means for systolic blood pressure, diastolic blood pressure, and pulse rate.

3.6.4 A Step-by-Step Guide to Comparing More than Two Vectors of Means Using the Excel Add-In

1. You must have a column in Excel that contains the names by which your cases are known. These are called the "case identifiers". They may be names or codes which you can use to identify the different cases, or may be simply case numbers (e.g. case 1, case 2, etc.). You must also have columns of data in Excel containing the variables whose means you want to include in the comparison and a column that tells Excel which group each case is in.
2. Go through the multivariate analysis add-in's menus until you get the dialogue box for multivariate tests of significance.
3. In the "Variables to use in analysis:" box, put the range of cells corresponding to the variable you are using in the analysis.
4. In the "Group identifiers:" box, put the range of cells corresponding to the column that indicates which group each case in the dataset is in.
5. Make sure the Yes/No choice for "Variable names in first line of data?" is appropriate for the ranges you have entered at Steps 3 and 4.
6. Make sure "Means" is selected for "Compare vectors of means or covariance matrices?".
7. Click "OK".

The analysis should now take place. The results will be shown in a new workbook in Excel.

3.7 COMPARING MORE THAN TWO COVARIANCE MATRICES

3.7.1 What Are We Testing, and How?

If you have read the previous Sections 3.4, 3.5 and 3.6, you can probably guess everything that will be written in this section. When comparing two covariance matrices in Section 3.5, we first transformed the data by taking absolute deviations from the group medians and then applying Hotelling's

T^2 test (the test used to compare two vectors of means in Section 3.4). Here we are going to employ a similar approach: transforming the data by taking absolute deviations from the group medians and then applying the likelihood ratio test that we used to compare more than two vectors of means in Section 3.6.

Yet again we consider the three variables used in Sections 3.4, 3.5, and 3.6 from the dataset described in Chapter 1: systolic blood pressure, diastolic blood pressure and pulse rate. We consider the four smoking habit groups as in Section 3.6 but rather than compare mean vectors, we are now comparing covariance matrices for the four groups. The hypotheses we examine are as follows.

- H_0: groups defined by smoking history have the same covariance matrices for systolic blood pressure, diastolic blood pressure and pulse rate.
- H_1: H_0 is not true.

3.7.2 Assumptions Made

For the final time in this chapter, let us consider the assumptions that we need to make before bothering to carry out all the necessary calculations for the test. They are essentially the same as for the three previous tests discussed in this chapter, as follows.

1. The cases in the data are independent of each other.
2. The data come from a multivariate normal distribution.
   - That is, the absolute deviations from the medians need to follow normal distributions.
3. The covariance matrices for the different populations being investigated are the same.
   - As we remarked upon when comparing two covariance matrices in Section 3.5, it feels slightly odd to have to make an assumption about the different covariance matrices being the same when we want to compare different covariance matrices to see if they are the same. However, as before, the assumption is about the covariance matrices of the absolute deviations from the group medians rather than the original variables. This assumption can best be examined by looking at the different covariance matrices involved and seeing whether or not they look similar.

3.7.3 Combining Levene's Method and the Likelihood Ratio Test

Before undertaking the Wilks' lambda test of Section 3.6, we need to transform the data we are using in the same manner as when carrying out the multivariate Levene's test in Section 3.5. That is, for each group's data, we deduct the median for each variable. We could use the means instead of the medians, but for reasons discussed in Section 3.5, using the medians gives a test that is more robust if our assumption of multivariate normality is a bit doubtful.

Having transformed the data by taking away the group medians for each variable, we get the following four vectors of means for the smoking habit groups:

$$\overline{\text{abs. diff}}_{\text{never smoked}} = \begin{pmatrix} 8.35 \\ 4.15 \\ 7.95 \end{pmatrix} \qquad \overline{\text{abs. diff}}_{\text{occasional smoker}} = \begin{pmatrix} 4.75 \\ 3.69 \\ 5.81 \end{pmatrix}$$

$$\overline{\text{abs. diff}}_{\text{ex-smoker}} = \begin{pmatrix} 6.60 \\ 3.35 \\ 6.60 \end{pmatrix} \qquad \overline{\text{abs. diff}}_{\text{current smoker}} = \begin{pmatrix} 6.92 \\ 3.46 \\ 5.83 \end{pmatrix}$$

The distributions of these absolute deviations reveal that they are similar to normal distributions, so we can be happy with this assumption.

The four covariance matrices are listed below. Although there are some differences between them, they are not so great that we feel convinced we need to abandon the test at this point.

$$\mathbf{S}_{\text{abs. diff,never smoked}} = \begin{bmatrix} 44.438 & 5.767 & -7.264 \\ 5.767 & 9.464 & 1.521 \\ -7.264 & 1.521 & 32.921 \end{bmatrix}$$

$$\mathbf{S}_{\text{abs. diff,occasional smoker}} = \begin{bmatrix} 28.867 & -5.683 & -10.850 \\ -5.683 & 7.563 & 6.338 \\ -10.850 & 6.338 & 25.496 \end{bmatrix}$$

$$\mathbf{S}_{\text{abs. diff,ex-smoker}} = \begin{bmatrix} 17.516 & -1.221 & -5.326 \\ -1.221 & 3.503 & 2.200 \\ -5.326 & 2.200 & 26.779 \end{bmatrix}$$

$$\mathbf{S}_{\text{abs. diff,current smoker}} = \begin{bmatrix} 20.514 & -0.178 & 0.203 \\ -0.178 & 7.824 & 1.841 \\ 0.203 & 1.841 & 26.841 \end{bmatrix}$$

We calculate the $\mathbf{W}$ and $\mathbf{T}$ matrices and work out the determinants to be $|\mathbf{W}| = 5,599,990,843$ and $|\mathbf{T}| = 6,225,677,243$, giving $\Lambda = 0.899$. With the rest of the formula for φ, these yield a value of $\varphi = 10.539$ with $p(m-1) = 9$ degrees of freedom again. This gives a p-value of 0.309.

As the p-value is more than 5%, we conclude that we have insufficient evidence to reject the null hypothesis of Section 3.7.1 at the 5% level of significance. We thus believe that the covariance matrices for the four smoking habit groups are sufficiently similar as to be regarded as the same. This gives added justification for the assumption that was made in Section 3.6 when we were comparing the mean vectors means for the four groups.

3.7.4 A Step-by-Step Guide to Comparing More than Two Covariance Matrices Using the Excel Add-In

1. You must have a column in Excel that contains the names by which your cases are known. These are called the "case identifiers". They may be names or codes which you can use to identify the different cases, or may be simply case numbers (e.g. case 1, case 2, etc.). You must also have columns of data in Excel containing the variables whose covariance matrices you want to include in the comparison and a column that tells Excel which group each case is in.
2. Go through the multivariate analysis add-in's menus until you get the dialogue box for multivariate tests of significance.
3. In the "Variables to use in analysis:" box, put the range of cells corresponding to the variables you are using in the analysis.
4. In the "Group identifiers:" box, put the range of cells corresponding to the column that indicates which group each case in the dataset is in.

5. Make sure the Yes/No choice for "Variable names in first line of data?" is appropriate for the ranges you have entered at Steps 3 and 4.
6. Make sure "Covariance matrices" is selected for "Compare vectors of means or covariance matrices?".
7. Click "OK".

The analysis should now take place. The results will be shown in a new workbook in Excel.

3.8 MORE INFORMATION

More information about the topics addressed in this chapter can be found in a surprisingly smaller number of places than you might expect. Many books about multivariate analysis concentrate on topics discussed elsewhere in this book and do not attempt to address the issue of multivariate tests of significance at all. However, I can recommend the books by Manley (2005) and Morrison (2005).

Factor Analysis 4

4.1 WHY DO I WANT TO DO FACTOR ANALYSIS?

In many areas of research, the concepts of interest are not directly measurable. This is particularly true in the social sciences where intelligence, social class, etc. cannot be measured. Often the researcher will, instead, collect other data which may be indicators of the unmeasured variable. For example, IQ or educational attainment can be measured as an indicator for intelligence, and occupation is often used as an indicator of social class.

Factor analysis is designed for just this situation. A set of variables is taken and the interrelationships between them are analysed to see whether a relatively small number of underlying, unobservable factors give rise to the data collected. Factor analysis is similar to multiple regression except that the dependent variables are the observed variables, and the regressors are the unobserved factors.

Take, for example, the correlation matrix in Matrix 4.1. It is called a matrix but a correlation matrix is really just a table of correlations as shown in Matrix 4.2. The correlation of a variable with itself is automatically 1 so that is what is appearing in the top left to bottom right diagonal in Matrix 4.1. It is also symmetric with the correlation between V1 and V2 being the same as the correlation between V2 and V1.

In Matrix 4.1, five variables have been recorded. The first two variables are very highly correlated with each other but not with the other variables. The last three variables are also very highly correlated with each other but not with the first two variables. It thus looks like we have a situation where there are two underlying factors at work. The first factor is what is behind the results for the first two variables: we might suppose that if someone has a high score for this unknown factor, then this will show itself through high values for both the first and second variables that we know about and have been measured. Similarly, there is a second factor which is behind the results for the last three variables.

MATRIX 4.1   A Hypothetical Correlation Matrix

$$\begin{bmatrix} 1\cdot00 & 0\cdot90 & 0\cdot05 & 0\cdot05 & 0\cdot05 \\ 0\cdot90 & 1\cdot00 & 0\cdot05 & 0\cdot05 & 0\cdot05 \\ 0\cdot05 & 0\cdot05 & 1\cdot00 & 0\cdot90 & 0\cdot90 \\ 0\cdot05 & 0\cdot05 & 0\cdot90 & 1\cdot00 & 0\cdot90 \\ 0\cdot05 & 0\cdot05 & 0\cdot90 & 0\cdot90 & 1\cdot00 \end{bmatrix}$$

MATRIX 4.2   Structure of a Correlation Matrix

$$\begin{bmatrix} \text{correlation between} & \text{correlation between} & \text{correlation between} \\ \text{V1 and itself} & \text{V1 and V2} & \text{V1 and V3} \\ \text{correlation between} & \text{correlation between} & \text{correlation between} \\ \text{V2 and V1} & \text{V2 and itself} & \text{V2 and V3} \\ \text{correlation between} & \text{correlation between} & \text{correlation between} \\ \text{V3 and V1} & \text{V3 and V2} & \text{V3 and itself} \end{bmatrix}$$

But what are these two factors? Because they have not been measured themselves, we do not really know. However, we can guess at what sort of thing they are by looking at what variables they are behind. For instance, say that the first variable in Matrix 4.1 was the result of a mathematics test involving addition, and the second variable was the result of a test involving multiplication. We might then suppose on the basis of these two variables that the first factor had something to do with the ability of a person to do mathematics. Similarly, the third variable in Matrix 4.1 might be the result of a reading test, the fourth variable the result of a spelling test and the fifth variable the result of a reading comprehension test. We might then suggest that the second factor which underlies these three variables has something to do with the ability of a person with language.

There is a further issue concerning how these two underlying factors are related to each other. In Matrix 4.1, the first two variables have very low correlations with the last three variables, and this suggests that the mathematics and language factors are therefore not related to each other. Now, it is not sensible for me to speculate in this book about whether or not this sounds reasonable. On the basis of the made-up example in Matrix 4.1, it is the case, and we shall leave that argument to one side. However, this issue does raise an important point, which is that the factors obtained in factor analysis are usually assumed to be independent of each other. This is a constraint that may not be appropriate in some situations, but fortunately there is a way around this. In Section 4.7 we look at factor rotation, and although the most common form of factor rotation preserves the independence of factors, there are also methods that allow factors to be correlated after rotation has taken place.

There are two reasons why you might want to undertake a factor analysis. One is to explore the data and try to identify any underlying factors that bring about the data observed. This is termed *exploratory factor analysis.* Alternatively, you might already have an idea of what factors exist behind the data which are being measured. You might then be conducting an analysis to try to confirm that the data observed match with this idea. This is called *confirmatory factor analysis.* Both exploratory and confirmatory factor analysis can be carried out using the methods described in this chapter. The main difference is in how one draws conclusions. The example used here concerns exploratory factor analysis but for confirmatory factor analysis, one would go through exactly the same decisions and thought processes except for at the end, where one would compare what has been discovered with what is said by the theory that is being tested.

4.2 WHAT DATA DO I NEED FOR FACTOR ANALYSIS?

For factor analysis, we need to have correlations between variables with which to work. For continuous data (or data which can be treated as continuous), the standard Pearson correlations can be calculated. For ordinal data (categorical data where the categories can be put in a meaningful order), Spearman's rank correlations can be calculated. See Chapter 1 for a discussion of types of data.

A correlation matrix can be calculated for a dataset with only a handful of cases. However, the more cases that exist in a dataset, the more reliable the correlations will generally be because of the additional information about the relationships between the variables that exists. There are lots of suggestions as to how much data is needed for a factor analysis to be reliable. I will not attempt to give a detailed review of all these suggestions, but merely list some below. They are not to be treated as firm rules that must be adhered to, but should instead be considered guidelines. Some appear to be almost contradicting others, and you should only be concerned if your dataset is distinctly smaller than the minimum they suggest.

- You should have ten times more cases in your dataset than the number of factors you wish to interpret.

- You should have five times more cases in your dataset than the number of variables in your dataset.
- You should have at least 100/200 cases in your dataset (take your pick!).

4.3 THE REST OF THIS CHAPTER

As the factors underlying the observed data are by their very nature unobservable, you will not be too surprised to hear that undertaking a factor analysis to understand more about them is not an exact science. There are a variety of ways of *extracting* the factors, and Section 4.4 deals with two of the more common methods. Once we have an idea about what the factors are, we can construct scores that we reckon the cases in our dataset might have if we were able to measure these unmeasureable factors. This is dealt with in Section 4.11. In Section 4.7 the issue of factor rotation is addressed. This is essentially finding a way of displaying the solution that is easier to interpret. Section 4.12 gives a step-by-step guide to undertaking a factor analysis. The chapter finishes with Section 4.13, where we give sources of more information about factor analysis.

4.4 HOW DO WE EXTRACT THE FACTORS?

To find out what factors underlie the observed data, the only clues we have are the relationships between the observed variables. We thus need to look at the correlations between all the variables. We could look at the covariances but these are scale dependent. This means that if we changed our measurement units (e.g. from centimetres to metres), the covariances would change in magnitude. However, the correlations do not change just because the measurement scale changes and so they are normally preferred for factor analysis.

Because the factors are unknown, we have to impose some sort of constraint on what they might be like in order for us to get anywhere with the analysis. Even in confirmatory factor analysis where we might have some idea about what the factors might be, it is best practice at this stage to treat them as unknown. One of these constraints is that we are able to express our observed variables as linear combinations of the factors, as in Equation 4.1, the general factor analysis model:

$$x_1 = \lambda_{11} f_1 + \lambda_{12} f_2 + \cdots + \lambda_{1k} f_k + u_1$$

$$x_2 = \lambda_{21} f_1 + \lambda_{22} f_2 + \cdots + \lambda_{2k} f_k + u_2$$

$$\vdots$$

$$x_p = \lambda_{p1} f_1 + \lambda_{p2} f_2 + \cdots + \lambda_{pk} f_k + u_p$$

(4.1)

In Equation 4.1, the x_1, x_2, ..., x_p are the p observed variables and the $f_1, f_2, \ldots,$ f_k are the k unobserved factors. The λs are like regression coefficients, and the $u_1, u_2, \ldots, u_p$ are like the error terms in regression equations. In factor analysis language, the λs are called *loadings*.

So we have a regression situation and can easily work out the coefficients, λ, right? Well no – although this looks like regression, in reality we know nothing about the right-hand side of the equations. We know neither the λs nor the values of the factors or the error terms. It is because of this difficulty that a number of ways of proceeding exist. These are called *extraction methods*. Two of the most common methods are principal components analysis and principal axis factoring, and these are discussed in Sections 4.4.1 and 4.4.2, respectively. Which is the better method to use? I am afraid that this is not a question that can be answered. The two methods are simply different ways of trying to accomplish the same ends, and discussion as to which is better is rather pointless. The issue of which solution should be believed is addressed in Section 4.10. However, at this point it is worthwhile mentioning that on some occasions, an extraction method may fail. Many of them operate on the basis of coming up with a solution using an iterative method. At some point in the iterative process, an impossible solution may be obtained, and thus the extraction process may fail. If this is the case, you then become very grateful that there are a number of alternative extraction methods available!

4.4.1 Using Principal Components Analysis

Principal components analysis (PCA) is sometimes treated as a separate topic in its own right, outside the topic of factor analysis. Its aim is to take a multivariate dataset and reduce its *dimensionality*. That is, rather than have *p* variables in a dataset, PCA manipulates the data into *p components*. These are just linear combinations of the original p variables as in Equation 4.2. The coefficients a_{ij} are just simple multipliers like regression coefficients.

$$z_1 = a_{11}x_1 + a_{12}x_2 + \cdots + a_{1p}x_p$$

$$z_2 = a_{21}x_1 + a_{22}x_2 + \cdots + a_{2p}x_p$$

$$\vdots$$

$$z_p = a_{p1}x_1 + a_{p2}x_2 + \cdots + a_{pp}x_p$$

(4.2)

On the face of it, simply replacing the p original variables with p new components does not seem such a great advance. However, there are some characteristics of the components $z_1, z_2, ..., z_p$ that make them useful.

The first characteristic is that all the information that is contained in the variables $x_1, x_2, ..., x_p$ is still retained in the components $z_1, z_2, ..., z_p$, so that nothing has been lost in creating the components. The second characteristic is that there is an ordering to the components. Of all the $z_1, z_2, ..., z_p$, the first component, z_1, contains a greater amount of the information originally available through $x_1, x_2, ..., x_p$ than any of the other components. Similarly, z_2 contains more information than any of the components other than z_1, z_3 contains more information than any of the components except z_1 and z_2, and this pattern continues through all the components. This is very useful because it means that the bulk of the information in the original $x_1, x_2, ..., x_p$ is contained in the more important components, and some of the components can be discarded.

For instance, let us consider the dataset discussed in Chapter 1 and the scores from eighteen areas of general knowledge. From these eighteen variables we create eighteen components (I will get to the bit about how we do this later on – please stick with me for now). Having done this, we find that the first component can, on its own, account for over 38% of the information in the original eighteen variables. A list of the amount of information accounted for by each component can be seen in Table 4.1 (please do not worry about the column headed "Eigenvalue" at the moment – I will explain it later). This shows us that once we consider the first seven components, together they account for three-quarters of the information in the eighteen original variables (see the last column of the table). Depending on the needs of any analysis we are carrying out, we may then decide to disregard the last eleven components. This has reduced the *dimensionality* of the dataset from eighteen to seven components.

Table 4.1 also demonstrates another characteristic of the components. You will notice that the percentage of the information in the original data which they explain reaches 100% once the last, eighteenth, component is included. Now this may not seem too surprising to you but it does have an important implication. It means that each tiny bit of the information contained in the

TABLE 4.1   Percentage of Information Accounted for by Components

COMPONENT	EIGENVALUE	PERCENTAGE OF INFORMATION IN ORIGINAL 18 VARIABLES ACCOUNTED FOR BY COMPONENT	CUMULATIVE PERCENTAGE
1	6.847	38.040	38.040
2	1.797	9.981	48.021
3	1.471	8.172	56.194
4	1.182	6.566	62.760
5	0.908	5.046	67.806
6	0.854	4.745	72.551
7	0.705	3.917	76.467
8	0.639	3.547	80.015
9	0.580	3.224	83.239
10	0.529	2.937	86.175
11	0.453	2.514	88.690
12	0.401	2.227	90.916
13	0.378	2.098	93.015
14	0.343	1.908	94.922
15	0.299	1.661	96.583
16	0.237	1.314	97.897
17	0.221	1.229	99.126
18	0.157	0.874	100.000

original variables finds its way into just one of the principal components. There is no sharing of information between components, and this means that they are independent of each other. This has important implications for how we interpret the results of a factor analysis, and we will return to this later in Section 4.7.

How do we do this principal components analysis? What is actually happening is a mathematical transformation of the eighteen variables into eighteen components. This is often known as an *eigenanalysis*. Now, the details behind this mathematical method are beyond what I am trying to accomplish in this book. If you are interested in the details, I would suggest you look at one of the books mentioned in Section 4.13, but if you do so and are scared by what you see, please do not worry. The important thing to know is that this eigenanalysis produces things called *eigenvalues* and *eigenvectors*. The eigenvalues are things that tell us how much of the information in the original data each component accounts for. For our eighteen-variable dataset, they are shown in Table 4.1 and were used to calculate the last two columns of that table.

The eigenvectors are the a_{ij} coefficients of Equation 4.2 and tell us how each variable is related to each of the components. These are the things that we try to look at to get an idea of what each of the factors is about. We will get around to doing this in Section 4.5.

4.4.2 Using Principal Axis Factoring

There is an important conceptual difference between the approaches taken by PCA and principal axis factoring (PAF). If you look back at Equation 4.1, you will see that each variable is "explained" by factors f_1, f_2, etc., with λs as weights, but there is also a final error term associated with each variable. In PCA, these error terms are discarded, and the principal components are derived as straightforward mathematical rearrangements of the original variables (see Equation 4.2). PAF does not discard these error terms.

PAF uses a concept called a *reduced correlation matrix*. This is the original correlation matrix but instead of having ones on the diagonal, it has values called *communalities*. If we consider the general factor analysis model in Equation 4.1, we see that it suggests the variables are made up of contributions from two sources: the factors and the error term. Thus, any variation we see in a variable comes from two sources: the factors and the error term. The proportion of the variation due to the factors is called the *communality* because it comes from a common source: the factors. The variation due to the error term is called the *unique* or *specific variation* as each error term is particularly related to one variable only.

Thus, by replacing the ones on the diagonal of the correlation matrix with the communalities, PAF is saying that all it will try to do is explain the variation as expressed by the communalities. By contrast, PCA, by having ones on the diagonal, is trying to explain all the variation. It could thus be said that PAF is less ambitious in its aims but more realistic. By posing this as a negative and a positive, I am trying to show that it cannot easily be said that one method is better than another. They are simply two different ways of trying to tackle the same problem.

Some readers may have already spotted a problem with PAF. If we want to replace the ones on the diagonal of the correlation matrix with the communalities, then we obviously need to know what these communalities are. But in order to find out what the communalities are, we need to know how much of the variation is due to the factors, and we do not yet know what the factors are. To overcome this, an iterative procedure is used. If we can find an initial guess for what the communalities could be, then we could use this as a starting point. Going back to basics, we recall that a communality is the proportion of the variation for a variable that can be explained by the factors. Given that the

MATRIX 4.3 Hypothetical Reduced Correlation Matrix

$$
\begin{bmatrix}
0.85 & 0.90 & 0.05 & 0.05 & 0.05 \\
0.90 & 0.87 & 0.05 & 0.05 & 0.05 \\
0.05 & 0.05 & 0.88 & 0.90 & 0.90 \\
0.05 & 0.05 & 0.90 & 0.75 & 0.90 \\
0.05 & 0.05 & 0.90 & 0.90 & 0.84
\end{bmatrix}
$$

factors are supposed to bring about the observed variables, why do we not start by considering how much of the variation in a variable can be explained by the other variables in the dataset? To do this, we can undertake a straightforward multiple regression. For the variable x_1, we undertake a regression where all the other variables $x_2, x_3, ..., x_p$ are explanatory variables. We can then see how much of the variation in x_1 is explained by the regression by considering the *coefficient of determination* or R^2 value. This can be our initial guess of the communality for x_1. We repeat the process for all the other variables and thus obtain an initial guess of the communality for each. The hypothetical correlation matrix of Matrix 4.1 then looks like that shown in Matrix 4.3. The main diagonal elements have been replaced by communalities while the correlations in the off-diagonal positions have remained the same.

A PCA factor analysis of the reduced correlation matrix is then undertaken. However, this is not the end of the story because the communalities used so far have just been initial guesses. The results of the PCA factor analysis on the reduced correlation matrix are now taken and new communalities are calculated. For a particular variable, this is done by taking the eigenvectors resulting from the PCA factor analysis, squaring all the elements relating to this variable and adding them. Strictly speaking, it is elements of scaled eigenvectors (such as shown in Table 4.2) which are squared and summed. If this explanation sounds rather brief, then I apologise, but it is beyond the scope of this book to start showing the steps of the calculations in too much detail. I hope that it gets across the idea of what goes on rather than the detail. Readers who are still frustrated could look at one of the books mentioned in Section 4.13.

The new communalities are then put into the reduced correlation matrix to replace the initial guesses, and the whole process is repeated. In fact, it is repeated and repeated until eventually the changes between each iteration are so small that they are not important. We then have a final PAF solution that we can then go on to interpret (see Section 4.7).

There is one issue which I have carefully ignored in the above description of how the PAF method works. Once we have done one lot of PCA on the reduced correlation matrix and replaced the initial guessed communalities

TABLE 4.2 PCA Loadings for General Knowledge Area Data

GENERAL KNOWLEDGE AREA	COMPONENT					
	1	2	3	4	...	18
History of science	0.603	0.019	0.024	−0.339		−0.098
Politics	0.729	0.263	0.169	−0.234		−0.013
Sport	0.573	0.343	0.328	0.313		0.070
History	0.740	0.366	0.024	−0.267		−0.044
Classical music	0.435	−0.078	−0.752	−0.085		−0.052
Art	0.600	−0.359	0.061	−0.289		0.155
Literature	0.623	0.108	−0.369	−0.253		0.027
General science	0.663	−0.076	0.015	0.330		0.104
Geography	0.559	0.252	−0.387	0.408		−0.053
Cookery	0.501	−0.492	−0.291	0.206		0.078
Medicine	0.444	−0.662	−0.174	0.190		−0.112
Games	0.584	0.297	0.040	0.428		−0.035
Discovery and exploration	0.744	0.238	−0.140	−0.137		0.195
Biology	0.728	−0.022	0.206	0.233		−0.079
Film	0.665	−0.245	0.151	−0.216		0.007
Fashion	0.605	−0.380	0.373	−0.102		−0.137
Finance	0.738	0.271	0.089	−0.056		−0.103
Popular music	0.403	−0.394	0.360	0.137		0.081

with ones calculated from the scaled eigenvectors, the next lot of communalities calculated will be identical to those just calculated if we use all the factors extracted by the PCA of the reduced correlation matrix. This is because using all the factors extracted by a PCA does, by the nature of its mathematics, completely reproduce what it started with. However, we know from the eigenvalues (such as shown in Table 4.1) that not all the factors extracted by the PCA will actually account for much of the information in the dataset, and many can be considered to be no more than noise. By excluding these noise factors from the calculation of the new communalities in PAF, we overcome this problem of the iterating procedure getting stuck because the PCA solution without the noise factors will not exactly reproduce the same communalities again and again. So what are the noise factors? This is something we tackle in Section 4.6, but for now let us rest our investigation of PAF until Section 4.7 with the knowledge that it can be carried out if we know how many factors we want to include in our final solution.

4.5 INTERPRETING THE RESULTS OF A PCA FACTOR ANALYSIS

Let us return to our principal components analysis. The eigenvalues of the correlation matrix are shown in Table 4.1. If we want to try to identify what the underlying factors might be, we need to look at the eigenvectors (also known as *loadings*) created in the eigenanalysis. They are shown in Table 4.2. First, just take a look at the columns of numbers in this table and see if there is a pattern. If you look closely, you will spot that the loadings for the first component tend to be larger (further from zero – in a positive or negative direction) than those for the second component, which in turn are a bit larger than those for the third component, and so on. There is a good reason for this: the eigenvectors have been displayed in such a way that the importance of the component has been taken into account. The first component is the most important, so its loadings are larger than the others. Similarly, the second component is more important than those that follow it, and so on. The loadings for the eighteenth component are all very near zero.

We now try and work out from Table 4.2 what the underlying factors might be. If we look at the first component, we see that a lot of the loadings are relatively similar in size, in the range 0.5 to 0.75. Only three of them are below 0.5: those for Classical music, Medicine and Popular music. Is it a coincidence that both of the music-related general knowledge areas are in this group? Probably not. We might interpret this component as revealing that the first (and therefore most important) factor in determining the scores is a person's knowledge about "things in general", excluding music and medicine. Some people simply know lots about "things in general". It is not a skill that can be easily measured but the results of the general knowledge tests and our subsequent use of factor analysis has revealed it as an important factor.

What about medicine and music? Why are they excluded? Well, if we now look at the second component, we see that the area which is most important is, in fact, Medicine. The fact that the loading is negative is not significant when it comes to importance. It is the fact that it is not near zero that makes it important. This implies that someone's knowledge of medicine is a factor in its own right. OK, so there are some other areas whose loadings are not an awfully long way away from the loading for Medicine, but this interpreting of factors is not an exact science – we need to be able to talk in generalities, and generally speaking, the second factor is much more closely connected to someone's knowledge of medicine than anything else.

By the time you get to this paragraph, some of you may already have an idea of what to do next, and have spotted that in the third component, there is again just one area which has a high loading: Classical music. Thus, the third factor is apparently associated with how much knowledge of classical music a person has.

The fourth factor is more difficult to interpret. High loadings are observed for Geography and Games, with loadings for History of science, Sport and General science not far behind. However, the loadings are not really that high. This indicates that this is a less important factor than the previous ones. Also of great importance is the fact that although the loadings for Geography, Games, Sport and General science are all positive, the loading for History of science is negative. This has an impact on how we interpret the factor. We could say that the fourth factor is thus associated with how *great* a knowledge of geography, games, sport and general science a person has AND how *little* knowledge they have of the history of science.

We could go on and on but we would find that it would become more and more difficult to make sensible interpretations. This is hardly surprising because as we go through the components, we find ourselves dealing with underlying factors that are increasingly unimportant. Once we reach the last component, there is nothing of note at all. What do we conclude? We conclude that it is just random noise and does not really mean anything.

4.6 HOW MANY FACTORS ARE THERE?

In Section 4.5 we saw that the interpretation of the results of the factor analysis was not too bad when we were examining the first few components/factors. However, when we came to the less important factors, things started to get difficult; and by the time we were at the last component/factor, we simply regarded it as noise. This raises the question about when we should we stop trying to interpret factors. I am afraid to say that there is no simple answer. There are a number of suggested rules, but none are perfect.

One suggestion is that we should only bother looking at factors if the eigenvalue associated with it is greater than one. Thus, in Table 4.1 we see that we have four eigenvalues greater than one, and should therefore try and interpret just four factors and disregard the rest. However, the fifth eigenvalue is 0.908, and thus the fifth factor explains only slightly less than the fourth factor. It could then also be argued that the sixth eigenvalue is only slightly smaller than the fifth, and we should therefore try and interpret the sixth factor as well. It would be more difficult to continue on with this reasoning to the seventh factor as the seventh eigenvalue is noticeably smaller than the sixth factor. Anyway, the above argument shows that having an arbitrary cut-off of one is sometimes difficult to justify. So why is it mentioned? The reason for using one as the suggested cut-off is that it is the average eigenvalue. The sum of the eigenvalues always equals the number of variables (eighteen in this case), so the average is exactly one.

Another suggestion to determine the number of factors is to say that we want to have enough factors so that cumulatively they account for at least 80% of the information in the original variables. In our example, Table 4.1 shows that we would need eight factors. Of course, there is nothing special about the 80% cut-off, so we might quite justifiably choose 70% (six factors) or 75% (seven factors) instead.

A third suggestion is something called a scree plot. This is a simple plot of the sizes of the eigenvalues against their order. A scree plot for the eigenvalues in our example (in Table 4.1) is shown in Figure 4.1. Now, a scree plot is so called because it is trying to identify the rubbish at the bottom of a cliff-face (which is called scree). At least a cliff-face is what we would like our scree plot to look like. If our scree plot looked like Figure 4.2, then we could say that the first three eigenvalues are at the top of the cliff and the remaining sixteen are the rubbish at the bottom of the cliff. However, all we can say from our real scree plot in Figure 4.1 is that the first factor is definitely important, and beyond that we are unsure!

So, how many factors should we interpret for the dataset we are examining here? The answer is, "It's up to you!" So long as you can defend what you are doing should someone challenge you, then you are at liberty to go along with any of the above ways of deciding how many factors you should have.

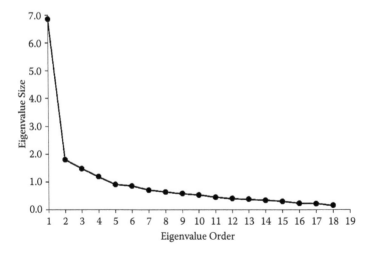

FIGURE 4.1 Scree plot for eigenvalues of correlation matrix for general knowledge data.

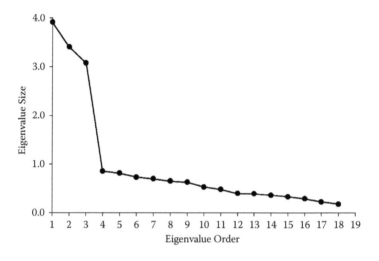

FIGURE 4.2   Ideal scree plot.

For some readers, this may be an uncomfortable message, but I am afraid you are stuck with it.

It is tempting for me to finish this section here and not actually commit myself to saying what I would do when faced with the general knowledge dataset. However, this would be unfair to readers and, anyway, I need to make a decision before I go on to Section 4.7. I am going to choose six factors to extract because although only the first four eigenvalues are above one, the fifth and sixth are not too far behind. There is then a bit of a drop to the seventh eigenvalue, and thus I am going to regard this as part of the noise and not related to a real factor. Also, including the fifth and sixth factors increases the amount of information accounted for from just 63% with four factors to over 70% (see cumulative percentage column of Table 4.1). It would be nice for this figure to be higher, but we will have to live with this relatively low figure.

4.7 INTERPRETING THE RESULTS OF A PAF FACTOR ANALYSIS

As explained in Section 4.4.2, we need to decide how many factors we are going to extract before undertaking a principal axis factoring (PAF) factor analysis. As explained in my concluding remarks in Section 4.6, I have decided that six factors are appropriate here.

The first thing to be aware of when interpreting a PAF factor analysis is that the eigenvalues associated with the final solution are not the same eigenvalues that are associated with the correlation matrix. This is hardly surprising as the PAF eigenvalues are obtained from a matrix which is different from the correlation matrix (recall Section 4.4.2). Examining Table 4.3, we see that now the five factors cumulatively only account for about 59% of the variation, which is considerably less than the 73% that Table 4.1 suggested might be obtained. The eigenvalues for Factors 5 and 6 have become much lower than one, indicating that they might not be so important after all. Even the eigenvalue for Factor 4 has descended to well below one. This suggests that maybe we should only concern ourselves with three factors, and the PAF factor analysis rerun with this information used when calculating the communalities at each iteration. However, this would make very little difference to the figures in Tables 4.3 and 4.4, so we continue our interpretation of the output shown in these tables.

The pattern of loadings shown in Table 4.4 is quite similar to that for PCA. For the first factor, there are three low loadings that stand out: for Classical music, Medicine and Popular music. It could be argued that the loading for Cookery is now also low enough to be classed with these, but the interpretation of this first factor is essentially unchanged: it is how much people know about "things in general" apart from Music and Medicine (and possibly Cookery).

Just as with the PCA factor analysis, the second factor is undoubtedly connected to knowledge of Medicine, and the third factor is connected with Classical music. For the fourth factor, not much stands out: mainly games and geography. Here, the solution differs slightly from the PCA solution which had other variables coming into this factor as well. The fifth factor is rather tricky to interpret, and I will leave this to one side as it is one of the least important factors extracted and quite possibly should be regarded as just noise. Rather

TABLE 4.3   Percentage of Information Accounted for by Five Factors Using PAF

FACTOR	EIGENVALUE	PERCENTAGE OF INFORMATION IN ORIGINAL 18 VARIABLES ACCOUNTED FOR BY 6 EXTRACTED FACTORS	CUMULATIVE PERCENTAGE
1	6.459	35.881	35.881
2	1.384	7.688	43.569
3	1.094	6.077	49.646
4	0.778	4.322	53.968
5	0.478	2.658	56.626
6	0.446	2.477	59.103

TABLE 4.4   PAF Loadings for General Knowledge Area Data

GENERAL KNOWLEDGE AREA	FACTOR					
	1	2	3	4	5	6
History of science	0.573	−0.012	0.007	−0.229	−0.282	0.105
Politics	0.714	0.248	0.120	−0.223	−0.158	−0.031
Sport	0.558	0.311	0.267	0.236	−0.126	−0.267
History	0.734	0.356	−0.009	−0.251	−0.066	0.090
Classical music	0.425	−0.103	−0.685	−0.053	0.104	−0.065
Art	0.568	−0.294	0.058	−0.181	0.060	0.073
Literature	0.598	0.075	−0.309	−0.186	0.147	−0.132
General science	0.639	−0.076	0.018	0.263	−0.221	0.173
Geography	0.538	0.198	−0.323	0.310	−0.070	−0.143
Cookery	0.477	−0.424	−0.205	0.125	−0.152	−0.149
Medicine	0.430	−0.603	−0.123	0.157	−0.121	0.130
Games	0.566	0.243	0.038	0.363	0.249	−0.089
Discovery and exploration	0.725	0.194	−0.127	−0.092	0.109	0.146
Biology	0.732	−0.033	0.230	0.288	0.210	0.339
Film	0.643	−0.219	0.132	−0.183	0.258	−0.015
Fashion	0.597	−0.363	0.366	−0.139	0.149	−0.256
Finance	0.713	0.236	0.051	−0.060	−0.081	0.017
Popular music	0.372	−0.273	0.232	0.026	−0.112	−0.167

conveniently though, the sixth factor is more easily interpreted, with Biology having the highest loading. However, this loading is not very large. We will return to the issue of how easy it is to interpret factors in Section 4.9 when we consider rotating factor solutions.

4.8 COMMUNALITIES BRIEFLY REVISITED

In Section 4.4.2 we explained about communalities. For a particular variable, the communality represents the proportion of its variation that can be accounted for by the factors extracted. For a PCA factor analysis where

TABLE 4.5   Final Communalities for General Knowledge Area Data

	COMMUNALITIES	
GENERAL KNOWLEDGE AREA	PCA FACTOR ANALYSIS	PAF FACTOR ANALYSIS
History of science	0.763	0.471
Politics	0.709	0.661
Sport	0.685	0.623
History	0.764	0.741
Classical music	0.791	0.678
Art	0.654	0.454
Literature	0.716	0.532
General science	0.714	0.563
Geography	0.735	0.554
Cookery	0.655	0.511
Medicine	0.757	0.620
Games	0.773	0.583
Discovery and exploration	0.683	0.622
Biology	0.765	0.832
Film	0.724	0.579
Fashion	0.740	0.730
Finance	0.635	0.578
Popular music	0.795	0.308

all the factors are considered worth extracting, the communalities will be exactly one for all the variables. However, this never happens in reality, so communalities associated with a final solution will be less than one. For the PCA factor analysis and PAF factor analysis of the general knowledge data, with six factors extracted, the final communalities are displayed in Table 4.5.

You will notice from Table 4.5 that the communalities from the PAF factor analysis are smaller than those from the PCA factor analysis. This is not surprising as in the PCA factor analysis, the first five factors manage to account for about 73% of the variation in the data, whereas the first five factors in the PAF factor analysis only manage to account for 59% of the variation.

It can also be seen that there is considerable variation between the variables in the size of their communalities. There is also considerable variation between the extraction methods. Popular music has the highest communality for the PCA analysis but the smallest for the PAF analysis.

This highlights one "feature" of factor analysis: the method of analysis you choose can have an effect on the results you obtain. We return to this issue in Section 4.10.

4.9 ROTATING FACTOR LOADINGS

Consider for a moment a hologram, such as you may well have on a bank card. If you look at the hologram from one angle, you get one view of the picture. However, if you move the card around, then you see the picture from a different angle, and see a different view of whatever is pictured.

What does this have to do with factor analysis? Well, the results of extracting factors using a principal components analysis or principal axis factoring method give us one view of the solution. If we imagine the results as a hologram, we could perhaps look at the solution from a different angle and see a different view of the solution. If we could do this, it might well be possible that we would find a view of the solution that actually gives us a "better" view. In terms of interpreting the results of a factor analysis, a "better" view would be one which can be more easily interpreted.

This idea of looking at a factor analysis solution from a different angle is called *rotation*. If we go back to the hologram on your bank card, hold it in front of you and imagine three axes: one pointing straight up, one pointing right and one pointing straight at you. Now, when you move the card around to see a different view of the hologram, you can imagine the axes moving around, or *rotating*.

Another way of thinking about rotation is to imagine the televising of a football match. At certain points in the football match, different cameras are used to give different views of the action. Depending on the action, some cameras give a better view of what is occurring than others although all the cameras are recording the same action. What we want to do is find the angle that shows us the action best.

So far I have used the words "better" and "best" about the results of rotating a factor analysis solution, and just mentioned that this means a more interpretable solution. However, if we want to actually implement a rotation, we have to know how to do it mathematically. That is, for instance, we want to know exactly where to move the axes when you are looking at your bank card's hologram. So what makes one factor analysis solution more interpretable than another one? Let us return briefly to Sections 4.5 and 4.7. When interpreting these solutions, the easiest factors to interpret were probably Factors 2 and 3. Factor 2 was related to mainly just one of the general knowledge areas: Medicine, and Factor 3 was similarly related to just one area: Classical music. This then gives us a clue as to what makes a factor interpretable: it is very useful if a factor has high loadings (in absolute value terms, ignoring minus signs) for a small number of variables and low loadings (near zero) for the rest of the variables. So, can we come up with a way of rotating a solution so that the rotated solution has these characteristics? The answer is, 'We can try'. What we want to do is make the loadings for a factor as different from each other as possible and, from a mathematical point of view, this can be likened to making the variance as large as possible. This is the basis behind the *varimax* rotation. This is, by far, the most commonly used of all rotations. The exact mathematics behind it is beyond the scope of this book but the principles involved are those just outlined above.

The results of applying the varimax rotation to the PCA loadings of Table 4.2 are shown in Table 4.6. The results of applying varimax rotation to the PAF loadings of Table 4.4 are shown in Table 4.7. You should not let the fact that we call the columns "components" for PCA and "factors" for PAF worry you. It is merely a convention that is not very important.

For the PCA analysis, the rotated components are quite different from the unrotated ones. The first component has high loadings for History of science, Politics, History, Discovery and exploration, and Finance. Thus we might think of this component as being how "well-read" the subjects in the dataset are. That is, the component differentiates between people who have obtained information (about both past and current events) and those who have not. The second component has high loadings for Sport and Games which have a clear link, but there is also a high loading for Biology. It might be suggested that people who are interested in sport and games might also have a knowledge of how the human mind and body perform in such events and may thus know about biology as well. The third component is much easier to interpret: high loadings for Art, Film and Fashion indicate that this component has something to do with knowing about artistic trends. The high loadings for General science, Cookery and Medicine for the fourth component do not appear at first glance to have much to do with each other. Perhaps this component could be termed "general knowledge" as these three subject areas are things that everyone comes into contact with on a regular basis. The fifth component has high loadings for Classical music and Literature and is clearly related to arts culture (as opposed to the third component which might be thought of as more contemporary arts culture). The sixth factor is clearly related to knowledge of Popular music.

For the PAF analysis, the interpretation of the first three factors follows quite closely that of the PCA analysis. There is absolutely no difference in interpretation of Factor 1. In Factor 2, the PAF analysis does not have a high

TABLE 4.6　PCA Loadings for General Knowledge Area Data after Varimax Rotation

GENERAL KNOWLEDGE AREA	COMPONENT					
	1	2	3	4	5	6
History of science	0.792	−0.092	0.107	−0.333	−0.047	−0.050
Politics	0.730	0.270	0.238	0.018	−0.137	−0.165
Sport	0.398	0.635	−0.013	0.032	0.044	−0.348
History	0.744	0.291	0.205	0.071	−0.276	−0.058
Classical music	0.095	0.015	0.071	−0.275	−0.828	0.121
Art	0.316	0.009	0.649	−0.354	0.090	0.026
Literature	0.344	0.131	0.273	0.032	−0.701	−0.112
General science	0.413	0.435	0.015	−0.589	0.005	−0.081
Geography	0.210	0.542	−0.231	−0.227	−0.525	−0.130
Cookery	0.046	0.062	0.128	−0.650	−0.358	−0.285
Medicine	−0.020	0.023	0.300	−0.804	−0.114	−0.087
Games	0.074	0.830	0.217	−0.030	−0.173	0.031
Discovery and exploration	0.537	0.384	0.310	−0.098	−0.349	0.141
Biology	0.291	0.613	0.433	−0.337	0.051	0.023
Film	0.208	0.200	0.745	−0.107	−0.206	−0.179
Fashion	0.178	0.153	0.657	−0.170	−0.010	−0.473
Finance	0.611	0.412	0.193	−0.044	−0.190	−0.130
Popular music	0.104	0.056	0.187	−0.212	0.011	−0.837

loading for Biology and thus the factor is just related to knowledge of sport and games. In Factor 3, Film and Fashion are again important. However, Art is less important in the PAF analysis and at the same time, Popular music is more important. None of these differences, though, really change the basic interpretation of what the factor is about. For the fourth factor in the PAF analysis, we do see a complete disagreement with the PCA analysis. It is the fifth component in the PCA analysis which becomes the fourth in the PAF analysis, being associated with Classical music and Literature. This "promotion" of the factor is not a shock. When you consider the eigenvalues for the fourth and fifth components/factors, they are very similar, so the fact that they are swapped around is not too surprising. The fifth PAF factor is quite like the fourth PCA component, although the loading for Medicine is more dominant than for the PCA analysis. The PCA and PAF analyses completely

TABLE 4.7 PAF Loadings for General Knowledge Area Data after Varimax Rotation

GENERAL KNOWLEDGE AREA	FACTOR					
	1	2	3	4	5	6
History of science	0.596	0.029	0.170	−0.089	−0.279	0.018
Politics	0.710	0.276	0.256	−0.104	−0.050	0.050
Sport	0.357	0.669	0.199	0.078	−0.041	0.007
History	0.762	0.238	0.144	−0.230	0.029	0.168
Classical music	0.134	0.028	−0.016	−0.774	−0.245	0.027
Art	0.315	−0.044	0.446	−0.173	−0.287	0.203
Literature	0.371	0.165	0.229	−0.556	−0.038	0.066
General science	0.365	0.325	0.077	−0.043	−0.498	0.261
Geography	0.226	0.528	−0.085	−0.401	−0.230	0.062
Cookery	0.069	0.134	0.269	−0.294	−0.572	−0.047
Medicine	0.014	−0.055	0.260	−0.165	−0.701	0.174
Games	0.149	0.613	0.150	−0.201	−0.009	0.350
Discovery and exploration	0.549	0.229	0.154	−0.360	−0.070	0.332
Biology	0.310	0.346	0.273	−0.012	−0.276	0.683
Film	0.296	0.063	0.584	−0.222	−0.131	0.283
Fashion	0.192	0.153	0.789	−0.015	−0.202	0.080
Finance	0.601	0.347	0.187	−0.158	−0.096	0.164
Popular music	0.127	0.156	0.411	0.080	−0.302	−0.038

disagree on the sixth factor. Whereas it was Popular music for the PCA analysis, it is Biology for the PAF analysis. The reason for this may be due to the previous five factors. Whereas the information about Biology was incorporated into the second component in the PCA analysis, it is not accounted for by any of the first five PAF factors and therefore still needs to be accounted for in the sixth factor. At the same time, Popular music was never likely to be the sixth factor for the PAF analysis as it had already been incorporated into its third factor.

We have solutions from the PCA and PAF analyses which give us partially different interpretations (although there are many clear points of agreement as well). We will discuss how we overcome this slight difficulty in Section 4.10. One thing that is very clear though is that interpreting all six components/factors has been easier when looking at the rotated solutions than when looking at the unrotated solutions. We can thus see that the process of rotating the solutions has achieved its goal.

The amount of variation in the data accounted for by each rotated factor is different from that accounted for by the unrotated factors. The percentages for each component/factor are shown in Tables 4.8 and 4.9. If you compare these tables with Table 4.1 and Table 4.3, you will see that the six rotated components/factors cumulatively account for the same amount of information as the unrotated components/factors. The main difference is that the first rotated factor is not as dominant as the first unrotated factor and that, as a result, the other components/factors have become more important than before.

There are several other rotation methods that try to do things similar to the varimax rotation. It is beyond the scope of this book to discuss these but I must stress again the fact that, by far, the most common rotation method to use is varimax.

TABLE 4.8 Percentage of Information Accounted for by Varimax Rotated PCA Components

COMPONENT	EIGENVALUE	PERCENTAGE OF INFORMATION IN ORIGINAL 18 VARIABLES ACCOUNTED FOR BY COMPONENT	CUMULATIVE PERCENTAGE
1	3.155	17.527	17.527
2	2.521	14.003	31.529
3	2.161	12.005	43.534
4	1.998	11.098	54.633
5	1.936	10.755	65.388
6	1.289	7.163	72.551

TABLE 4.9 Percentage of Information Accounted for by Varimax Rotated PAF Factors

FACTOR	EIGENVALUE	PERCENTAGE OF INFORMATION IN ORIGINAL 18 VARIABLES ACCOUNTED FOR BY FACTOR	CUMULATIVE PERCENTAGE
1	2.933	16.296	16.296
2	1.736	9.645	25.941
3	1.847	10.263	36.204
4	1.542	8.567	44.771
5	1.585	8.807	53.578
6	0.994	5.525	59.103

4.9.1 Non-Orthogonal/Oblique Rotations

The results of using a principal components or principal axis factoring method for extracting the factors give us factors that are independent of each other. This means that there is no correlation between these underlying factors. Now, while that may well be a perfectly rational position to take in many circumstances, there may also be occasions on which this does not seem so realistic. In these circumstances, a type of rotation can be undertaken which gives factors that can be correlated with each other. These are called *non-orthogonal* or *oblique* rotations (whereas rotations where independence is maintained are called orthogonal). There are various sorts of rotation that give this result but one of the most commonly used is *promax*. Further coverage of this here would take this book beyond its scope of covering the essentials of multivariate analysis.

4.10 SO WHICH SOLUTION DO WE BELIEVE?

In the above sections we have seen four solutions: unrotated PCA, unrotated PAF, rotated PCA and rotated PAF. I have also mentioned that there are other extraction and rotation methods. Which should you believe?

The answer to this question is partly philosophical and partly practical. From a philosophical point of view, you either believe in rotating a solution or you do not. Most people who carry out factor analysis are happy to carry out rotation but there are some who see it as not being good practice. They argue that just because a solution is not easily interpreted, this is not a good reason to throw it away and find something that is easier. However, a counter-argument could be that so long as you have decided to rotate the solution before you begin the analysis, and stick to this decision no matter what the unrotated solution looks like, then you are not simply picking the solution that fits your preconceptions best. So my advice would be to make a decision about rotating and stick to it. If you have decided to rotate, then do not bother looking at the unrotated solution. Similarly, if you have decided not to rotate, then go no further than interpreting the unrotated solution.

That philosophical part of the answer helps us cut down the number of solutions to either rotated or unrotated ones. We still have a number of possible extraction methods available; and although in this chapter our example dataset gives very similar solutions whether PCA or PAF methods are used, this is not necessarily always going to be the case. If you are faced with competing solutions that are sufficiently different from each other so as to be confusing, then you need to come up with some means of getting a majority decision made. That is, if you can do a number of analyses, you may well find that a lot give similar solutions, and this can then be what you decide to choose as your final solution. How might we get several solutions? One way is to use a variety of extraction methods. Another is to carry out your analysis on different subsets of your data. That is, randomly divide your dataset into a number of smaller datasets and carry out factor analyses on each part. You may then find that you get similar solutions being given on several occasions, and this can help you make a decision on what solution is most reliable. Of course, you need a large enough dataset in the first place to use this tactic. Ultimately, if your analyses cannot yield a solution that is obtained on several occasions, then you have to be honest and simply report that you have solutions that compete with each other.

4.11 FACTOR SCORES

Once factors have been identified, *factor scores* can be created. These are measures of the supposed unmeasureable factors and are obtained by treating the loadings in the factor analysis solution as kind of regression coefficients. Because the variables in the dataset are correlated with each other, the loadings are initially scaled by the correlation matrix (technically speaking, the matrix of loadings is pre-multiplied by the inverse of the correlation matrix, but please do not worry about this if you are not so familiar with matrix calculations). This process gives coefficients which are used to multiply the standardised original data. The original data are standardised (for each value, the mean of the variable is subtracted and the result is divided by the variable's standard deviation) because they may be measured on different scales, and not standardising would mean that variables with larger values (in general) would have an influence on the factor scores that was more than would be appropriate.

The factor scores which are obtained can be used to examine the cases in the dataset and, for instance, rank them according to the levels of the factor scores. Sometimes factor scores are used in further analyses (such as regressions) as if they were variables in their own right. If you want to do this, I should issue a caution here. In the further analyses, it must be remembered that the factor scores are not something which have been carefully measured, and a different factor analysis solution would give difference factor scores. There is thus a good degree of uncertainty in the factor scores that should be remembered when subsequently using them.

4.12 A STEP-BY-STEP GUIDE TO FACTOR ANALYSIS USING THE EXCEL ADD-IN

1. You must have a column in Excel that contains the names by which your cases are known. These are called the "case identifiers". They may be names or codes which you can use to identify the different cases, or may be simply case numbers (e.g. case 1, case 2, etc.). You must also have columns of data in Excel containing the variables which you want to use in the factor analysis.
2. Go through the multivariate analysis add-in's menus until you get the dialogue box for factor analysis.
3. In the "Case identifiers:" box, put the range of cells corresponding to the column in which the case names, labels or whatever (see Step 1) are located.
4. In the "Variables to use in analysis:" box, put the range of cells corresponding to the variables you are using in the analysis.
5. Make sure the Yes/No choice for "Variable names in first line of data?" is appropriate for the ranges you have entered at Steps 3 and 4.
6. Put nothing in the "Number of factors to extract:" box at this stage and click "OK".
7. From the results of the eigenanalysis that are produced, decide how many clusters you want in your factor analysis solution (see Section 4.6).
8. Return to your original dataset and go through Steps 2, 3, 4 and 5 again.
9. Now, in the "Number of factors to extract:" box, put the number you decided on at Step 7.
10. Make sure that the extraction method chosen matches with what you want to use.
11. If you are happy to interpret a rotated solution, make sure that "Varimax" is selected for "Rotation method:". If not, make sure "None" is chosen.
12. Click "OK".
13. Interpret the output you get. If you want to try another analysis, perhaps using a different extraction method, return to Step 8.

4.13 MORE INFORMATION

There are many different books that have been written about factor analysis over time. Some are general books on multivariate analysis and cover factor analysis well [see, for instance, Afifi et al. (2012), Bartholomew et al. (2008) and Everitt and Dunn (2001)]. Others are books about factor analysis alone and are able to go into great depth about the subject. One that I can recommend is Fabrigar and Wegener (2012).

Cluster Analysis

5

5.1 WHY DO I WANT TO DO CLUSTER ANALYSIS?

If you want to investigate whether or not there are groups of cases in your dataset and what the characteristics of these groups are, then cluster analysis is the tool for you. It may be that you reckon that there are groups, and you may even have a good idea about how these groups are made up. In this case, you can do a cluster analysis to see if you are right. Alternatively, you may have very little idea about what groups exist and may then want to use cluster analysis as an exploratory process.

There are three main ways of approaching cluster analysis. Hierarchical clustering is the most commonly used and is thus the one discussed in this book. The others are non-hierarchical clustering and model-based clustering. Non-hierarchical clustering used to be more popular in the past because it is less computationally demanding for a computer than hierarchical clustering. These days, advances in computing make it possible to undertake hierarchical clustering in a mere fraction of the time that it used to take. Non-hierarchical clustering has thus fallen out of favour to some extent, although it is still very useful for enormous datasets where computing time still matters even today. We discuss non-hierarchical clustering briefly in Section 5.10. Model-based clustering is a much more technical affair. It works by fitting different models to the data and seeing which one fits best. I will not say more than that as it is beyond the scope of this book. However, those interested in pursuing this idea might want to look, in the first instance, at Everitt et al. (2011).

The idea of cluster analysis is to find "natural" groups of cases that exist in a dataset. I put the word *natural* in quotation marks because the word makes it sound like we are investigating biological or physical processes. Of course, if your data does involve this, then the word is quite appropriate – cluster analysis looks for naturally occurring clusters that exist out there in the world (or at least in the world as far as your dataset goes). In other contexts, the word "natural" is a little less obvious to use but essentially we are still talking about the same thing – it is clusters of cases that do actually exist in the dataset that are (hopefully) being revealed by the cluster analysis.

I indicated above that we would concentrate on hierarchical cluster analysis in this chapter. A typical cluster analysis of this type consists of three stages:

1. Starting with each case in the dataset as a cluster all on its own, join the two clusters that are closest to each other. Then with the new setup of clusters that results, join two of these clusters that are closest to each other. Continue doing this until all the cases are joined up in one cluster. Figure 5.1 gives a graphical representation of this stage, starting at the bottom with seven individual cases which gradually join together until they are all in one cluster at the top.
2. Decide how many "natural" clusters exist and identify which cases are in which cluster.
3. Interpret the characteristics of the clusters.

These three stages are addressed in this chapter. Stage 1 occupies Sections 5.4 and 5.5. Stage 2 is discussed in Section 5.6 and Stage 3 in Section 5.8.

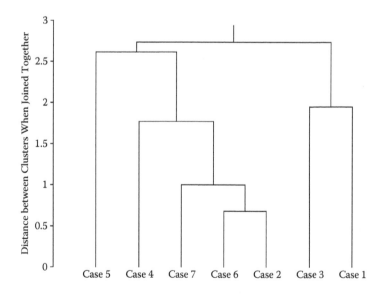

FIGURE 5.1 Graphical representation of Stage 1 of hierarchical cluster analysis.

5.2 WHAT DATA DO I NEED FOR CLUSTER ANALYSIS?

For cluster analysis, continuous data (or data which can be treated as continuous) is required. See Chapter 1 for a discussion of types of data. You can use binary data as if it were continuous but there are, in fact, some special adaptations of cluster analysis possible for binary data. This is mentioned a bit more in Section 5.4.

5.3 THE REST OF THIS CHAPTER

We continue this chapter by discussing the issue of how we measure the distance between two cases in a dataset (Section 5.4) and between two clusters (Section 5.5). Having given you various options in these two sections, I try in Section 5.6 to help you decide what to choose. Once these issues are resolved, the issue of how many clusters exist in the data is addressed in Section 5.7. Interpreting the clusters is discussed in Section 5.8. To conclude the chapter we have a section outlining non-hierarchical clustering (Section 5.9) before giving a step-by-step guide to undertaking a cluster analysis using the Microsoft® Excel® add-in that accompanies this book (Section 5.10) and sources of further information about cluster analysis (Section 5.11).

5.4 HOW DO WE DECIDE HOW CLOSE TOGETHER TWO CASES ARE?

In Figure 5.1 we can see that when cases or clusters join together, there is also an axis showing how far there was between the cases or clusters when they were joined together. In this section I show you some ways of defining distances between cases. Linked with this is the idea of distances between clusters where there are multiple cases in the clusters. This is discussed in Section 5.5.

Let us consider the dataset discussed in Chapter 1 and look at the variables systolic blood pressure, diastolic blood pressure and pulse rate to see if we can find any groups in the dataset. However, before considering all 100 cases in

TABLE 5.1   Data for First Three Cases in Dataset

CASE NUMBER	SYSTOLIC BLOOD PRESSURE (mm Hg)	DIASTOLIC BLOOD PRESSURE (mm Hg)	RESTING PULSE RATE (BEATS PER MINUTE)
1	94	63	82
2	130	70	68
3	111	66	71

the dataset, let us look in more detail at just the first three cases. Their data are listed in Table 5.1.

There are a number of ways of defining the distance between the cases. The following Sections 5.4.1 through 5.4.6 explain some of the more common methods. A decision must be made when undertaking a cluster analysis as to which distance measure is going to be used. This question is discussed in Section 5.6.

5.4.1 Distances by Absolute Value

How far is Case 1 from Case 2? Well, we could say that for systolic blood pressure, they are thirty-six units apart; for diastolic blood pressure, they are seven units apart; and for pulse rate, they are fourteen units apart. Put all these together and we conclude that they are fifty-seven units apart (36 + 7 + 14). Similarly, we can conclude that Cases 1 and 3 are thirty-one units apart (17 + 3 + 11) and that Cases 2 and 3 are twenty-six units apart (19 + 4 + 3). This is a nice intuitive way of thinking of distances. Mathematically, we are taking *absolute values*. That is, we are taking the values of one case away from another and ignoring minus signs before adding them. The distance between Cases 1 and 2 on systolic blood pressure can be written as 94 − 130 = −36. For diastolic blood pressure, the difference is 63 − 70 = −7 and for pulse rate is 82 − 68 = 14. If we just added these, we would have some negatives cancelling out positives and get a distance of −29. If we take absolute values and ignore the minus signs, we get 36 + 7 + 14 = 57. It is important that we do this. If not, we could get negatives cancelling out positives completely in some cases. We would then say that two cases had no distance between them when their data could be quite different.

5.4.2 Standardising

Despite having been clever enough to be taking absolute values in Section 5.4.1, we are still making an important error when calculating the distances. In the dataset as a whole, the systolic blood pressure has a mean of 111.36 and

TABLE 5.2   Standardised Data for First Three Cases in Dataset

CASE NUMBER	SYSTOLIC BLOOD PRESSURE (mm Hg)	DIASTOLIC BLOOD PRESSURE (mm Hg)	RESTING PULSE RATE (BEATS PER MINUTE)
1	−1.714	−0.094	1.283
2	1.840	1.064	−0.195
3	−0.036	0.402	0.121

a standard deviation of 10.13. What this tells us is that a change of one unit in the systolic blood pressure is a change of less than 10% in its standard deviation (1/10.13 = 9.87%). However, a one-unit change in diastolic blood pressure (mean = 63.57; standard deviation = 6.04) is a change of over 16.5% of its standard deviation (1/6.04 = 16.56%). In Section 5.4.1 we were treating a change of one unit in systolic blood pressure as being of the same importance as a one-unit change in diastolic blood pressure. This cannot be a good thing to do.

The solution to this problem is to work with standardised data rather than the raw data of Table 5.1. In Table 5.2 we see the standardised data for the first three cases in the dataset. To standardise the data, I have deducted the mean and divided by the standard deviation. That is, the 94 from Table 5.1 for Case 1's systolic blood pressure has been standardised by deducting the mean for systolic blood pressure (111.36, giving 94 − 111.36 = −17.36) and dividing by the standard deviation (10.13, giving −17.36/10.13 = −1.714). The other values have been standardised in the same way.

5.4.3 Distances by Absolute Value Using Standardised Data

With the standardised data, the distance between Cases 1 and 2 using absolute values is 6.190. This comes from −1.714 minus 1.840, giving −3.554 which is 3.554 in absolute value; −0.094 minus 1.064, giving −1.158 which is 1.158 in absolute value; and 1.283 minus −0.195, giving 1.478. The 6.190 comes from summing the absolute values: 3.554 + 1.158 + 1.478.

The distance between Cases 1 and 3 by the same method is 3.336 and that between Cases 2 and 3 is 2.854.

5.4.4 Euclidean Distances

I am aware that by using the name of an ancient Greek mathematician (Euclid) in the title of this section, I may have caused a bit of panic in some readers.

As we will see, there is really nothing more natural than this distance measure. I will explain how we work out Euclidean distances first and then explain why they are so natural.

Having learnt the lesson of Section 5.4.2, we will use standardised data. When we calculated distances using absolute values in Sections 5.4.1 and 5.4.3, the reason for using absolute values was so that negative differences did not cancel out positive differences. Another way of accomplishing the same task is to square the differences instead. This always gets rid of negative numbers. Having removed the negative numbers by squaring the differences, we add up these squared values and finally take the square root to get rid of all that squaring that we have been doing.

For example, from Table 5.2, what is the Euclidean distance between Cases 1 and 2? We start with a difference of −3.554 on the systolic blood pressure (−1.714 − 1.840 = −3.554) and square it to get 12.631. Similarly, the difference for diastolic blood pressure is −1.158 (−0.094 − 1.064 = −1.158), which we square to get 1.341. Lastly, the difference for pulse rate is 1.478 (1.283) − (−0.195) = 1.478), which is 2.184 when it is squared. Adding these squared values gives 16.156. The Euclidean distance is then obtained by taking the square root of this and getting 4.019.

The distance between Cases 1 and 3 by the same method is 2.100 and between Cases 2 and 3 is 2.014.

Hopefully having reached this paragraph, you now realise that despite a Greek mathematician being involved, it is not too difficult to calculate Euclidean distances. But now I should try and convince you that they are also quite natural. In order to do so, I am going to have to involve another Greek mathematician, but you are quite likely to have heard of this one: Pythagoras. In Figure 5.2 we see a right-angled triangle. Those of you who remember what you might have learnt about Pythagoras' theorem will recall that the length of the hypotenuse (side A to C) can be calculated by squaring the difference between A and C horizontally (the side of the triangle A to B), squaring the difference between A and C vertically (the side of the triangle B to C), adding

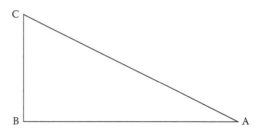

FIGURE 5.2 A right-angled triangle.

these together and taking the square root. This is exactly what we just did to work out the Euclidean distance: we worked out the difference between Cases 1 and 2 according to one direction and squared it (the systolic blood pressure can be thought of as a "direction" in multivariate data). We did the same for the other two directions we had (diastolic blood pressure and pulse rate) and summed the squared values before finally taking the square root. Thus, Euclidean distance is just what we have in our world as natural distances. In one, two or three dimensions, we can measure distances with a ruler, and these distances are Euclidean distances. The reason we call this Euclidean distance rather than Pythagorean distance is because Euclid went beyond the work of Pythagoras into many dimensions. So, no matter how many variables we want to work with to calculate distances, Euclidean distances can be calculated.

5.4.5 Squared Euclidean Distances

You might be able to guess what I am going to explain in this section just by examining its title. Squared Euclidean distances are just Euclidean distances which have been squared. Or rather, from a computational point of view, they are Euclidean distances where we have not bothered to take the square root after adding all the squared bits.

Why bother with squared Euclidean distances? One good reason is that they are often used, but a stronger reason is that sometimes you might want to exaggerate distances. Two Euclidean distances of 2 and 4, for instance, would be squared Euclidean distances of 4 and 16. Using this distance measure rather than ordinary Euclidean would have the effect of discouraging the joining of two clusters that have any cases some distance apart, even if most of the cases in the two clusters are near each other. It produces a different pattern to the joining up of cases and clusters. Whether this is a better pattern or a worse pattern is not really a question that can be answered. We return to this issue in Section 5.6.

5.4.6 Distances for Binary Data

Most cluster analyses involve continuous data or data that can be treated as continuous, and the distance measures discussed above are suitable for these sort of data. It is possible to use binary data (where the values are either zero or one) with these methods, but it is not ideal. For a particular binary variable, two cases either have the same value or they do not. This is a different concept from continuous data when the idea of "how near" naturally gives rise to distances such as those discussed above.

Where all the variables in a dataset are binary, an alternative approach is to ask what proportion of the variables do two cases have the same values for. This gives us an idea of how similar two cases are. For instance, if two cases have the same values for seven out of ten binary variables, then they have a *similarity* of 70%. This may also be viewed as a *dissimilarity* of 30%. This dissimilarity can be viewed as a type of distance. The more differences exist between two cases, the larger the dissimilarity will be, just like a distance measure for continuous data.

There are a number of variations on this idea that have been suggested over the years. It is beyond the scope of this book to go into the details of these but interested readers might look at books mentioned in Section 5.11.

5.5 HOW DO WE DECIDE HOW CLOSE TOGETHER TWO CLUSTERS ARE?

Section 5.4 has given us some ideas as to what distance measures we might want to use when considering how close two cases are to each other. However, how do we go about defining the distance between a case and a cluster, or between two clusters? There are many different possibilities or *linkage methods* to choose from. Some of the most commonly used are shown below. When undertaking a cluster analysis, a decision as to which linkage method to use must be made. This question is discussed in Section 5.6.

5.5.1 Average Linkage between Groups

Let us start by thinking how we might define the distance between a single case and a cluster of cases. Figure 5.3 illustrates the position we are thinking about.

When considering how far Case D is from the Cluster A/B/C, we have three individual distance measures involved: A to D, B to D, C to D. These will be defined by whatever distance measure it has been decided will be used for the analysis. Perhaps the most intuitive way of defining the distance between the case and the cluster is to use the average of these three distances. This method is *average linkage between groups*.

When considering the distance between two clusters, we have a situation like Figure 5.4. There are now six distances between the two clusters: A to D, A to E, B to D, B to E, C to D, C to E. The average linkage between groups

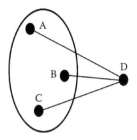

FIGURE 5.3 A case and a cluster.

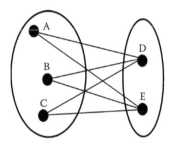

FIGURE 5.4 Two clusters.

method takes an average of all six of these distances to measure the distance between the two clusters.

5.5.2 Complete Linkage

For *complete linkage*, rather than taking averages, the largest distances are used. Thus for Figure 5.3, the distance between the case and the cluster would be the largest of A to D, B to D and C to D. For Figure 5.4, the distance between the two clusters would be the largest of A to D, A to E, B to D, B to E, C to D and C to E.

By using just one distance rather than all, complete linkage ignores most of the distances, and in particular ignores small distances. Thus, although two clusters might have cases which are relatively close to each other, the distances between the clusters will be determined only by cases which are far away from each other. So, if you judge that you want to have clusters containing cases which are as tightly bunched together as possible, this may be the method for you.

5.5.3 Single Linkage

This linkage method can be regarded as the opposite of complete linkage. Instead of taking the largest distances from the selection available, *single linkage* takes the smallest. For Figure 5.3, single linkage chooses the smallest of the distances A to D, B to D and C to D. For Figure 5.4, it chooses the smallest of A to D, A to E, B to D, B to E, C to D and C to E.

In contrast to complete linkage, it disregards all distances apart from the smallest ones. Thus, although some of the cases in one cluster may be a considerable distance from some cases in another cluster, all it takes is one case from each cluster to be near each other for the distance between the clusters to be regarded as small. You might choose to use this method if one of the most important criteria for you is that cases which are near to each other are in the same cluster, regardless of the fact that some cases in the cluster may be a long way from each other.

5.5.4 Forward-Thinking Linkage Methods

In Sections 5.5.1 through 5.5.3, we have seen linkage methods which calculate distances between clusters based on the situation that exists at one point in the process of turning separate cases (at the bottom of Figure 5.1) into clusters of cases and eventually into one large cluster (at the top of Figure 5.1). There are other linkage methods worth mentioning which do not work on this principle, but instead consider what *would be* the case if two clusters were joined.

In Figure 5.5 we see the state of a hierarchical clustering at some point in the clustering process between the bottom of something like Figure 5.1 and the top of something like Figure 5.1. The linkage methods in Sections 5.5.1 through 5.5.3 would calculate all the distances between the two clusters and Case D (A/B/C to D, A/B/C to E/F, D to E/F), and whichever gave the smallest result would be the next step in the movement from the bottom to the top of a figure like Figure 5.1.

By contrast, forward-thinking linkage methods consider the three possible scenarios that could occur (see 1, 2, 3 below). They look at the clusters which are formed if the proposed joining together from 1, 2 or 3 takes place and assesses how compact the clusters are that are formed.

1. A/B/C joins with D. This leaves us with two clusters (A/B/C/D and E/F).

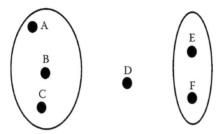

FIGURE 5.5 Two clusters and a case.

2. A/B/C joins with E/F. This leaves us with one cluster (A/B/C/E/F) and one case (D).
3. D joins with E/F. This leaves us with two clusters (A/B/C and D/E/F).

Let us take possibility 1. We have a cluster A/B/C/D. How good is A/B/C/D as a cluster of cases? One way of measuring this is to look at all the distances between cases in this cluster. That is, A to B, A to C, A to D, B to C, B to D, C to D. Taking an average of these would be what the forward-thinking linkage method *average linkage within groups* does. The smaller this average, the closer together the cases are in the cluster. The larger the average, the more spaced out the cases are. We then define the distance from A/B/C to D as the average we have just calculated from the cluster that would be formed if we put them together. We do the same for possibility 2 (A/B/C joining with E/F) and possibility 3 (D joining with E/F). The possibility that we decide to implement is the one that has given us the smallest distance.

Another forward-thinking method looks not just at the new cluster that would be formed, but at all the other clusters as well. Going back to possibility 1, we have our new cluster A/B/C/D but also our existing cluster E/F. The linkage method called *Ward's method* looks at how far each case is from the centre of its cluster. That is, for the cluster A/B/C/D, we calculate the centre (that is, the means of the variables, calculated from the cases in the cluster) and calculate the squared Euclidean distance of each of A, B, C and D to this centre. We also do the same for the cluster E/F, finding its centre and then calculating the squared Euclidean distance from E and F to the cluster centre. We add up all the squared Euclidean distances obtained from A, B, C, D, E and F, and define that as being the resulting distance measure associated with going ahead with possibility 1.

We then carry out the same sort of calculations for possibilities 2 and 3, and end up with a resulting distance measure for each. We then see which of possibilities 1, 2 and 3 give the smallest distance measure and implement that as the next step in going from the bottom to the top of something like Figure 5.1.

5.6 HOW DO WE DECIDE WHICH DISTANCE MEASURE AND LINKAGE METHOD TO USE?

I suppose the answer to this question is that you should use whichever distance measure and linkage method is most appropriate to the aims of your analysis. So, for example, if the exaggeration of normal distances caused by using the squared Euclidean distance measure is appropriate (see Section 5.4.5), then you should use this. Also, for example, if it is appropriate to define distances between clusters along the lines that complete linkage uses (see Section 5.5.2), then this should be the linkage method you use.

However, you are more likely to be in a situation where you are not sure what distance measure and linkage method to use. There may be no way of deciding that some decisions are more appropriate than others for your analysis. How do you proceed in these circumstances? Well, you must recognise the situation you are in and that by choosing a distance measure and linkage method, you will get just one of a number of possible sets of results. This sounds bad but if you repeat the analysis using different distance measures and linkage methods, then you get a range of results and can see what the general consensus is regarding the number of clusters and the characteristics of these clusters. You will hardly ever get identical solutions from different distance measures and linkage methods, but hopefully you can spot common patterns. You may then choose one combination of distance measure and linkage method which produces results typical of the general consensus. When you finally report the results of the analysis, you can say that a range of cluster analyses gave results similar to the ones presented. This gives added weight to the conclusions you draw.

5.7 HOW DO WE DECIDE HOW MANY CLUSTERS THERE ARE?

5.7.1 Using First Seven Cases in the Dataset

The picture in Figure 5.1 is often called a *dendrogram*. It shows how the clustering of cases occurs with them starting off as individuals at the bottom and then joining up until they are all in one cluster at the top. The scale shows the distance between clusters when they are joined. Thus, in Figure 5.1, we see that Cases 2 and 6 join together at a distance of just under 0.75. Case 7 then joins the cluster containing Cases 2 and 6. There is then a bit of a jump in distances to about 1.75 when Case 4 joins up with Cases 2, 6 and 7. Cases 1 and 3 then join up with a distance of about 2. There is then another jump in distances until Case 5 joins with Cases 2, 4, 6 and 7 at a distance of about 2.6 and then shortly thereafter, these cases join with the cluster containing Cases 1 and 3. The original seven cases are now in just one cluster.

This clustering can be represented in a table, such as Table 5.3. You can reproduce this for yourself – it is a clustering of the first seven cases in the dataset of Chapter 1, with systolic blood pressure, diastolic blood pressure and pulse rate as the variables used. The data is standardised before use with the Euclidean distance measure, and average linkage between groups is the linkage method.

But how many clusters exist? The answer must be between one and seven inclusive, and the dendrogram can help us decide. If we look at Figure 5.6, we see the various possibilities marked on the dendrogram. If we were to decide on the seven cluster solution, then we would be saying that none of the cases were anything like each other and should be regarded as completely separate. This is rarely the case in reality, and here we have cases joining together at a relatively small distance (0.680), and so reject this seven cluster solution. We also reject the six cluster solution because shortly after going from seven clusters to six clusters, we then go to five clusters. However, to go from five clusters to four clusters, we need to increase our joining distance from 1.003 to 1.768. In terms of proportion, this is quite an increase and may indicate that we are joining two clusters which are quite different from each other. We can therefore construct an argument which says we have five clusters in our dataset. However, we should carry on in case we find an even more compelling argument for a different number of clusters. Once we are at four clusters,

TABLE 5.3 Table of Clustering Shown in Figure 5.1

STEP	JOINING DISTANCE	JOINING	JOINING	NEW CLUSTER CALLED
1	0.680	Case 2	Case 6	Case 2/Case 6
2	1.003	Case 2/Case 6	Case 7	Case 2, etc.
3	1.768	Case 2, etc.	Case 4	Case 2, etc.
4	1.948	Case 1	Case 3	Case 1/Case 3
5	2.614	Case 2, etc.	Case 5	Case 2, etc.
6	2.734	Case 1/Case 3	Case 2, etc.	

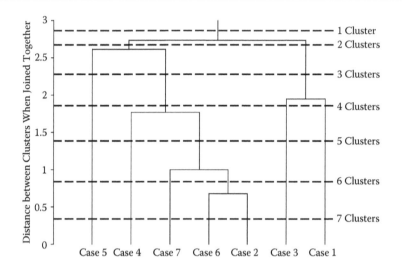

FIGURE 5.6   Dendrogram for first seven cases with possible numbers of clusters marked.

we go to three clusters quite quickly, and so reject four clusters as a possible solution. However, now we again note a sizeable jump if we wish to go to two clusters (from 1.948 to 2.614). Perhaps we should conclude that going from three clusters to two clusters is unwise and we should stick to three clusters. We thus have a second possible answer for 'How many cluster are there?'. Continuing on, we see that going from two clusters to one cluster involves only a small change in distance and thus there is no good reason to suggest that two clusters exist. Of course, we could decide that all the cases are so similar that only one cluster exists, but as we already have two possible solutions, we do not pursue this idea.

So, for Figure 5.1 and its tabular form Table 5.3, we have decided that there are either three or five clusters. Which do we choose? Well, we should look at our interpretations of both possibilities (see Section 5.8) and decide which seems more appropriate. It is entirely possible that we have to end up presenting both as possible solutions.

Additionally, we should really try out other distance measures and linkage methods as well (see Section 5.6). They may give different suggestions for the number of clusters. Even if they suggest the same number of clusters, we should remember that there is no guarantee that the same cases are grouped together in the different solutions. The real test is to interpret the results (see Section 5.8) and see if the same story is being told by the different methods, or at least try to find some consensus from the majority of methods tried.

5.7.2 Using All Cases in the Dataset

If instead of using just the first seven cases in the dataset we now use all 100 cases, we obtain the dendrogram of Figure 5.7. As before, the data are standardised before use with the Euclidean distance measure and average linkage between groups as the linkage method. It is much more difficult to assess from this dendrogram how many clusters we should choose. If we had many more cases (after all, 100 is not a very large dataset), the dendrogram would be completely useless.

So what do we do if we cannot use the dendrogram to assess how many clusters we have? The answer lies in doing exactly the same sort of thing as in Section 5.7.1. There, we considered the sizes of the jumps in distances, and we can do the same again now, using the tabular form of the dendrogram. For the dendrogram of Figure 5.7, we have Table 5.4. It does not show all the stages of the clustering, but only the final ten steps. Why only the last ten? Are we not interested in jumps in the joining distances at earlier steps? Well, these questions are very good ones, and the only answer is that of practicality. We could look at the joining distances and decide that there was a large jump between steps 40 and 41 for instance, but that would mean we would choose to have sixty clusters in the dataset. If we were to decide this, then the next step in the analysis would probably be throwing our arms in the air and giving up. Who can really try and interpret what sixty different clusters are telling us? Usually in cluster analysis, we are expecting a relatively small number of clusters to exist in the dataset, so it makes sense to concentrate on the later steps of the clustering procedure and look for jumps in distances there rather than at earlier steps.

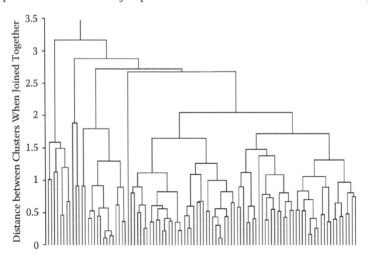

FIGURE 5.7 Dendrogram for all cases.

TABLE 5.4   Table of Clustering Shown in Figure 5.7

STEP	JOINING DISTANCE	JOINING	JOINING	NEW CLUSTER CALLED
⋮	⋮	⋮	⋮	⋮
90	1.587	Case "14", etc.	Case "20"/Case "21"	Case "14", etc.
91	1.654	Case "2", etc.	Case "12", etc.	Case "2", etc.
92	1.721	Case "1", etc.	Case "3", etc.	Case "1", etc.
93	1.795	Case "9", etc.	Case "18"/Case "25"	Case "9", etc.
94	1.883	Case "5"/Case "56"	Case "88"	Case "5", etc.
95	2.046	Case "1", etc.	Case "2", etc.	Case "1", etc.
96	2.680	Case "1", etc.	Case "38"	Case "1", etc.
97	2.723	Case "1", etc.	Case "9", etc.	Case "1", etc.
98	2.879	Case "1", etc.	Case "5", etc.	Case "1", etc.
99	3.167	Case "1", etc.	Case "14", etc.	

TABLE 5.5   Changes in Joining Distances from Table 5.4

STEP	JOINING DISTANCE	NUMBER OF CLUSTERS LEFT AFTER THIS STEP	INCREASE IN JOINING DISTANCE FROM PREVIOUS STEP	PERCENTAGE INCREASE IN JOINING DISTANCE
⋮	⋮	⋮	⋮	⋮
90	1.587	10	0.0918	6.14%
91	1.654	9	0.0664	4.18%
92	1.721	8	0.0674	4.07%
93	1.795	7	0.0740	4.30%
94	1.883	6	0.0881	4.91%
95	2.046	5	0.1631	8.66%
96	2.680	4	0.6333	30.95%
97	2.723	3	0.0433	1.62%
98	2.879	2	0.1561	5.73%
99	3.167	1	0.2882	10.01%

To help us look at the changes in joining distances shown in Table 5.4, we construct another table that is, Table 5.5. We now see that the increases in joining distances are relatively stable for steps 90 through to 94, but then almost doubles at step 95. The percentage change shows this even more dramatically. We could thus suggest that we should stop at step 94 and have six clusters in our results. However, it could also be suggested that we should examine one further step. The change in joining distance from step 95 to step 96 is large by comparison with those that precede it, and again the percentage increase shows this dramatically. Perhaps then we should choose a five cluster solution.

Of course, as mentioned in the final paragraph of Section 5.7.1, we should remember that other distance measures and linkage methods may give different results. Rather than pretend that other methods of doing the clustering do not exist, it is better to try a variety of methods and see if agreement can be reached by a number of these methods.

5.8 INTERPRETING CLUSTERS

Once you have decided how many clusters your analysis is showing you exist, you want to then go on and find out more about these clusters. A straightforward way of doing this is to calculate summary statistics for each cluster. Tables 5.6 and 5.7 show the number of cases in each of the six clusters that might be thought to exist as a result of the discussions of Section 5.7.2. Perhaps of most importance is Table 5.6 which shows the number of cases in each cluster and also the mean of the standardised values for the three variables involved in the clustering: systolic blood pressure, diastolic blood pressure and pulse rate. From examining this table, we can immediately see that "Cluster 1" contains thirty-nine people with lower than average systolic blood pressure (mean is negative), higher than average pulse rate (mean is positive) and about average diastolic blood pressure (mean is near zero). The other large cluster

TABLE 5.6 Summary Statistics Using Standardised Data for Six Cluster Solution of Section 5.7.2

	NUMBER OF CASES IN CLUSTER	MEAN BLOOD PRESSURE (SYSTOLIC)	MEAN BLOOD PRESSURE (DIASTOLIC)	MEAN PULSE RATE (RESTING)
Cluster 1	39	−0.461	0.092	0.801
Cluster 2	34	0.772	0.524	−0.071
Cluster 3	1	0.754	2.388	1.495
Cluster 4	15	0.076	−0.944	−1.618
Cluster 5	3	1.708	1.230	−1.357
Cluster 6	8	−1.911	−1.667	−0.248

TABLE 5.7   Summary Statistics Using Unstandardised Data for Six Cluster Solution of Section 5.7.2

	NUMBER OF CASES IN CLUSTER	MEAN BLOOD PRESSURE (SYSTOLIC)	MEAN BLOOD PRESSURE (DIASTOLIC)	MEAN PULSE RATE (RESTING)
Cluster 1	39	106.692	64.128	77.436
Cluster 2	34	119.176	66.735	69.176
Cluster 3	1	119.000	78.000	84.000
Cluster 4	15	112.133	57.867	54.533
Cluster 5	3	128.667	71.000	57.000
Cluster 6	8	92.000	53.500	67.500

is "Cluster 2", and it has cases with above average readings for both blood pressure variables. "Cluster 3" is an individual case on its own which has diastolic blood pressure and pulse rate readings that are a long way above average. "Cluster 4" is a group of fifteen cases which have average systolic blood pressure readings but have diastolic blood pressure and pulse rate readings well below average. "Cluster 5" has three cases that are way above average for the blood pressure readings but below average for pulse rate. The final cluster, "Cluster 6", has lower than average readings for all three variables.

Table 5.7 shows the same patterns but now the units involved are appropriate for the variable being considered. It is therefore less easy to see what are above and below average readings but, on a more positive note, the values in the table do have a substantive meaning.

Further investigations could be carried out trying to link things like gender, age and smoking history to the different clusters to see if any patterns emerged. The Excel add-in that accompanies this book produces information on which cluster each case in the dataset belongs to. This can then be cross-tabulated with these other variables just as you would produce any other cross-tabulation (e.g. age against smoking history). As this is fairly standard data manipulation, we do not go into further details here.

5.9 NON-HIERARCHICAL CLUSTER ANALYSIS

As explained in Section 5.1, non-hierarchical clustering is not used nearly as much as it used to be. Hierarchical clustering, discussed above, is far and away the most popular method of clustering. However, it is worth outlining here in brief the concept behind non-hierarchical clustering.

To start with, you need to decide how many clusters you intend to look for in your data before you start non-hierarchical clustering. This immediately makes it more awkward than hierarchical clustering unless you have some good reason for knowing this information. The general procedure for non-hierarchical clustering is then as follows:

1. Assuming that m clusters have been decided upon, the first step is to find m individual cases in the dataset which are as different from each other as possible. These act as initial "centres of gravity" for the clusters.
2. All the remaining cases are then put into one of the m clusters according to which of the m "centres of gravity" they are nearest.
3. For each of the m clusters, the "average" position is calculated. These average positions then act as new centres of gravity.
4. All the cases in the dataset are reallocated to one of the m clusters according to which of the new m centres of gravity they are nearest.
5. Steps 3 and 4 are repeated until each time the reallocation of cases is carried out, the same centres of gravity and allocations of cases to clusters are achieved.

5.10 A STEP-BY-STEP GUIDE TO CLUSTER ANALYSIS USING THE EXCEL ADD-IN

1. You must have a column in Excel that contains the names by which your cases are known. These are called the "case identifiers". They may be names or codes that you can use to identify the different cases, or may be simply case numbers (e.g. case 1, case 2, etc.). You must also have columns of data containing the variables which you want to use in the cluster analysis.
2. Go through the multivariate analysis add-in's menus until you get the dialogue box for cluster analysis.
3. In the "Case identifiers:" box, put the range of cells corresponding to the column in which the case names, labels or whatever (see Step 1) are located.
4. In the "Variables to use in analysis:" box, put the range of cells corresponding to the variables you are using in the analysis.

5. Make sure the Yes/No choice for "Variable names in first line of data?" is appropriate for the ranges you have entered at Steps 3 and 4.
6. Make sure the Yes/No choice for "Standardise data?" reflects whether or not you want to standardise the data before calculating distance measures.
7. There are three options for "Distance measure:" and three options for "Linkage method:" provided. Ensure that the combination you want to use is selected.
8. If you want to produce a dendrogram, make sure the "Yes" option is selected for "Produce dendrogram?".
9. If you want to investigate the characteristics of clusters, then put the number of clusters in the box for "Display cluster membership and summary statistics" and make sure the "Yes" option is selected.
10. Click "OK".

5.11 MORE INFORMATION

For more information about hierarchical and non-hierarchical clustering methods, there are few books which can approach that of Everitt et al. (2011). In this text, titled (not surprisingly) *Cluster Analysis*, the topic is discussed in some depth. However, other books that deal with the topic well are Bartholomew et al (2008) and (not surprisingly) Everitt and Dunn (2001).

Discriminant Analysis

6

6.1 WHY DO I WANT TO DO DISCRIMINANT ANALYSIS?

Discriminant analysis is all about what makes groups different from each other. You know that you have groups in your dataset but can you tell what group a case should belong to just from the variables you have? If you can, then you have variables that *discriminate* between the groups. But are all the variables useful for doing this, or just some? And how good are the variables at discriminating? Are they so great that they will always predict the correct group, or is there some uncertainty involved? Maybe for some cases you do not know the group to which they belong. Can information about cases whose group you do know help you allocate these other cases to groups?

All these questions and issues are addressed in discriminant analysis.

6.2 WHAT DATA DO I NEED FOR DISCRIMINANT ANALYSIS?

As with many of the topics in this book, you need to have data which is continuous, or at least can be regarded as continuous. See Chapter 1 for a discussion of types of data. You can also use binary data as if it were continuous. Of course, in addition to this, you need to have a variable which tells you about the groups you are investigating. There are no restrictions on the number of groups which can be handled, although you do need data for at least two cases from each group.

6.3 THE REST OF THIS CHAPTER

This chapter proceeds by initially discussing (in Section 6.4) the issue of deciding how close an individual case is to the different possible groups. This then leads to linear discriminant functions in Section 6.5 and how we might allocate individual cases to one of the possible groups. In Section 6.6 we tackle the issue of deciding which variables are useful for discriminating between groups. Section 6.7 discusses how accurate the allocation decisions are, and Section 6.8 deals with how we might judge how well a discriminant analysis performs. Section 6.9 briefly discusses some other methods of discriminant analysis before the chapter concludes, in Section 6.10, with directions as to how to use the Excel add-in which comes with this book to undertake a discriminant analysis and sources of more information on the topic in Section 6.11.

6.4 HOW DO WE DECIDE HOW CLOSE A CASE IS TO DIFFERENT GROUPS?

If we want to assess how well the variables we have can discriminate between groups, then we need to look at the characteristics of the groups in terms of the variables available.

Let us consider starting with an extremely trivial example. In the dataset discussed in Chapter 1, we have data on gender and height. We find that the mean height for males is 1.76 metres and for females is 1.61 metres. So, if someone else came along who had a height of 1.76 metres, we would guess that they belonged to the group "male"; and if they had a height of 1.61 metres, we would guess that the belonged to the group "female". Of course, it is not impossible for these allocations to be incorrect. It is perfectly possible for a male to have a height of 1.61 metres or a female to have a height of 1.76 metres. Just how accurate the allocation to the male or female group is likely to depend on how varied the height data are. We will return to this issue in Section 6.7.

I said the above example was trivial and indeed it was because we are really quite unlikely to have someone new turn up with a height of exactly 1.76 metres or exactly 1.61 metres. What is more likely is that they will have a height which is not exactly equal to one of the means for the two groups. So, what if the new person had a height of 1.69 metres? The intuitively obvious thing to do is to see how far away this measurement is from each of the group means we have. We find that this 1.69 metres is 8 centimetres away from the mean for the female group and 7 centimetres away from the mean for the male group. We would thus say that the new person is more likely to be male than female.

As a brief aside, the above decision to allocate the new person to the male group does make an assumption about the variability of the heights in the two groups. If we had an absurd situation where all the males were exactly 1.76 metres but the females could take a wide range of heights from, say, 1.2 metres up to 2.02 metres, then we might change our minds about the new person being most likely male. We would probably conclude that the person could be female as the height was within the range of other females for which we had data, but could not possibly be male as the height was not exactly 1.76 metres. We would thus allocate them to the female group. However, here, a more reasonable assumption would be that the variation in height amongst the female group would be similar to that amongst the male group. In this case, arguing that the 1.69 metre tall person was more likely to be male because their height was nearer to the mean for the male group than the mean for the female group is reasonable. We discuss this sort of assumption further in Section 6.4.2.

Of course, in reality we are going to be operating with more than one variable, and quite possibly with more than two groups. We thus have to come up with a way of measuring the distance between an individual case and a group on the basis of multivariate data. In Chapter 5 when we were dealing with cluster analysis, we discussed a number of possible distance measures. However, in discriminant analysis, there is another distance measure not covered in Chapter 5 which is commonly used. This is called the Mahalanobis distance (after the Indian statistician P. C. Mahalanobis). It is similar to the squared Euclidean distance using standardised data mentioned in Chapter 5, and in fact is identical to it if the variables being used have correlations between them of exactly zero. However, I can guarantee that you will never have a dataset in front of you where the variables have correlations of exactly zero, unless the dataset has been manufactured in some way to have this characteristic. Even where two variables are not associated with each other, simple random fluctuations will dictate that their correlation will not be exactly zero (although it may be quite close to zero).

The reason that the Mahalanobis distance is often used in discriminant analysis is because of the allowances it makes for correlations between the variables. Let us return for now to our example of two groups (males and females) and the variable height. We can now make this a multivariate example by adding two new variables: left foot length and right foot length. You are probably thinking, 'Isn't it a bit stupid to use both of these – won't they be almost the same?' Well, you are correct; you are indeed unlikely to do something like this in practice, but what is the precise reason? The answer you might give after a little thought would be that if you used both left foot length and right foot length, then you would be duplicating information. Their correlation will be near one, indicating the close relationship they have.

But what if instead of having left foot length and right foot length, we had systolic blood pressure and diastolic blood pressure? It does not seem silly to include both of these but they are certainly not independent of each other. Their correlation in the dataset of Chapter 1 is 0.590 – not as close to one as we might expect from left foot length and right foot length but nevertheless a correlation which is a long way from being zero. Does this not mean that we are duplicating information if we use both systolic and diastolic blood pressures in an analysis? The answer is, 'Yes'. The Mahalanobis distance overcomes this duplication of information by not just having a mechanism similar to dividing by the variance (thus achieving a standardising of the variables), but also by doing something similar to dividing by the covariance (which is closely related to the correlation – see Chapter 3). There may be some readers familiar with the workings of the Mahalanobis distance who are currently cringing at this description. I admit that a phrase such as "doing something similar to" is not often found in statistical books. However, for the purposes of this book, I am going to leave the explanation of the Mahalanobis distance at this stage. I cannot avoid showing the formula for the Mahalanobis distance in Section 6.4.1 (Equation 6.1), but a detailed knowledge of how the matrix calculations work is not necessary for someone who wants to use a straightforward discriminant analysis in their research.

Some of you who have read Chapter 5 on cluster analysis might be wondering why we are putting all this effort into avoiding the duplication of information in a discriminant analysis when we did not mention it in cluster analysis. The reason is the different emphasis placed on the variables in each type of analysis. In cluster analysis, the variables are not really important in their own right. We simply use them to try to identify clusters which are the objects of interest. In discriminant analysis, we already know what the groups involved are, so a lot of the attention is on the variables – which are the ones that are useful to discriminating between groups, and how good are they at discriminating? We could use the Mahalanobis distance as a distance measure in cluster analysis, but this is rarely done.

6.4.1 Linear Discriminant Functions

Although I am trying to avoid the use of too many formulae in this book, I am afraid that what is shown in Equation 6.1 cannot really be excluded. However, please do not worry about all the superscripts and subscripts – it is not necessary for you to understand the entirety of what is going on here but it is important that you, the reader, are aware, at least in part, of how we get from the Mahalanobis distance to the *linear discriminant functions*.

$$d^2_{\underline{x},\underline{m}_i} = (\underline{x} - \underline{m}_i)^T S_*^{-1}(\underline{x} - \underline{m}_i)$$

$$= \underline{x}^T S_*^{-1}\underline{x} - 2\underline{m}_i^T S_*^{-1}\underline{x} + m_i^T S_*^{-1}\underline{m}_i$$

$$= \underline{x}^T S_*^{-1}\underline{x} - 2[\underline{m}_i^T S_*^{-1}\underline{x} - 1/2 m_i^T S_*^{-1}\underline{m}_i] \qquad (6.1)$$

Independent of i Linear Function of $\underline{x}$ Constant for i

In Equation 6.1, the first line is the Mahalanobis distance between an individual case and a group. The values of the variables for the individual case are contained within the vector x and the averages of the variables over the cases known to be in group i are contained within the vector $\underline{m}_i$. If you are unsure about the word *vector*, please do not worry, and have a look at Chapter 3. The $(x - \underline{m}_i)$ indicates that for each variable, we are finding the difference between the value for the individual case and the mean for the group. The fact that $(x - \underline{m}_i)$ appears twice is because we are squaring the differences, just as when we calculate a variance or a Euclidean distance (see Chapter 5). The fact that one of the $(x - \underline{m}_i)$ terms has a "T" superscript should not concern you – it all has to do with matrix multiplication and is simply a way of writing it correctly so that the mathematics makes sense.

The S_* in the middle of Equation 6.1 is the pooled covariance matrix. That is, the covariance matrix for each group in the dataset has been calculated, and the weighted average of each cell in the matrix has been calculated. In Chapter 3 we calculated a pooled covariance matrix for two groups when carrying out Hotelling's T^2 test. Now we calculate a pooled covariance matrix for any number of groups in the same way:

$$S_* = \frac{(n_1 - 1)S_1 + (n_2 - 1)S_2 + \cdots + (n_g - 1)S_g}{(n_1 - 1) + (n_2 - 1) + \cdots + (n_g - 1)}$$

where we have g groups and n_1 cases in Group 1, n_2 in Group 2, etc. The fact that the S_* is raised to the power of -1 means that we are effectively dividing by S_*. In matrix mathematics, instead of dividing by a matrix, you multiply by its inverse. This is just the same as multiplying by 2^{-1} instead of dividing by 2 in normal mathematics. If you need reminding about what a covariance matrix is, please see Chapter 3.

The last two lines of Equation 6.1 show how the Mahalanobis distance can be amended so that it takes the form in the last line where there are three different parts making up the distance which is written as $d^2_{\underline{x},\underline{m}_i}$: the squared distance between the vector x and the vector of means, $\underline{m}_i$. This transformation of the Mahalanobis distance is generally attributed to the famous statistician, R. A. Fisher.

The key thing behind this change is that the three parts which make up the distance have difference characteristics. The first in Equation 6.1 only involves x, the data from the individual case and the pooled covariance matrix, S_*. Thus, if we are finding the distance between the individual case and Group 1 and also its distance from Groups 2, 3, etc., then this part of the distance only needs to be calculated once and then reused each time. Also, very importantly, if we want to find out which group the individual case is nearest, we can ignore it completely – all the distances between the case and the different groups have this component, so it tells us nothing about which group the individual case is nearest.

The final term in Equation 6.1 only involves $\underline{m}_i$ and the pooled covariance matrix, S_*. It thus needs to be calculated for the distance between the individual case and Group 1, and also for each of the individual case's distances to Group 2, 3, etc. This term does have an important role to play in deciding which group is nearest the individual case. However, because this term does not depend on the data for the individual case (there is no vector x in the term), then once we have calculated the term for the distances from one individual case to however many groups we have, we do not need to calculate it again if we happen to have more individual cases to examine.

The middle term in Equation 6.1 contains all of x, $\underline{m}_i$ and S_*. However, because $\underline{m}_i^T S_*^{-1}$ only needs to be calculated once (for each group), no matter how many individual cases we want to work with, and the vector x only appears once in the term, then we say that this term is a *linear function of x*. That is, it is like a regression equation with each of the variables that make up x having coefficients which are $\underline{m}_i^T S_*^{-1}$.

The part of the last line of Equation 6.1 in square brackets is what is commonly known as the linear discriminant function. In Section 6.5 we see it in use when we allocate individual cases to groups, and in Section 6.6 we see how it can be used to indicate how useful different variables are in discriminating between groups.

6.4.2 Assumptions Made

In a sense, we have not made any assumptions in creating the linear discriminant functions in Section 6.4.1 which are of the same importance as assumptions we were making in Chapter 3 when undertaking tests of significance. However, on the other hand, if we cannot assume that the covariance matrices for the different groups are sufficiently similar, then the action of pooling them to create the Mahalanobis distance makes little sense. It is not that the covariance matrices for the different groups have to be identical for us to be happy, but more that they should have figures which are at least of the same order of magnitude and sign (positive/negative).

The other thing that should be mentioned here is the fact that the Mahalanobis distance works best if the data have at least approximately a multivariate Normal distribution (see Chapter 3 for more details). This is not an essential requirement for the discriminant analysis to be valid, but if a simple transformation of the data can be done to achieve a better fit to Normality, then it is probably a good idea to do it.

6.5 ALLOCATING INDIVIDUAL CASES TO GROUPS

6.5.1 Creating the Linear Discriminant Functions

Let us consider the dataset of Chapter 1. One of the variables recorded is smoking history, and has four groups: "never smoked", "occasional smoker", "ex-smoker" and "current smoker". Using the variables age, weight, systolic blood pressure, diastolic blood pressure and pulse rate, let us construct the linear discriminant function for each group and then use them to allocate individual cases to the groups. Will the allocations be correct? Well, if the variables we are using do discriminate well between the groups, then we would expect a lot of correct allocations. However, if the variables have little to do with what groups the cases are in, then we can expect a lot of wrong allocations.

To construct the linear discriminant functions, we need to create the pooled covariance matrix. The individual covariance matrices for the four smoking history groups are as follows for the variables age, weight, systolic blood pressure, diastolic blood pressure and pulse rate.

$$
S_{\text{never smoked}} = \begin{bmatrix}
153.885 & 42.831 & -31.141 & -25.962 & -26.795 \\
42.831 & 210.964 & -7.744 & 6.478 & 36.763 \\
-31.141 & -7.744 & 112.438 & 31.044 & -39.482 \\
-25.962 & 6.478 & 31.044 & 27.126 & 9.856 \\
-26.795 & 36.763 & -39.482 & 9.856 & 97.241
\end{bmatrix}
$$

$$
S_{\text{occasional smoker}} = \begin{bmatrix}
174.600 & 53.467 & -17.133 & -26.717 & -66.483 \\
53.467 & 469.143 & 24.313 & -15.840 & -66.527 \\
-17.133 & 24.313 & 51.267 & 7.583 & -15.517 \\
-26.717 & -15.840 & 7.583 & 21.729 & 8.154 \\
-66.483 & -66.527 & -15.517 & 8.154 & 56.429
\end{bmatrix}
$$

$$S_{\text{ex-smoker}} = \begin{bmatrix} 177.516 & -64.769 & -43.400 & -35.789 & -41.663 \\ -64.769 & 331.837 & 25.199 & 18.764 & -2.514 \\ -43.400 & 25.199 & 61.589 & 6.763 & -26.463 \\ -35.789 & 18.764 & 6.763 & 15.250 & 15.263 \\ -41.663 & -2.514 & -26.463 & 15.263 & 71.958 \end{bmatrix}$$

$$S_{\text{current smoker}} = \begin{bmatrix} 115.520 & -62.603 & -6.920 & -16.929 & -37.942 \\ -62.603 & 299.812 & -47.666 & -5.792 & 33.641 \\ -6.920 & -47.666 & 70.406 & 15.152 & -11.159 \\ -16.929 & -5.792 & 15.152 & 19.897 & 6.435 \\ -37.942 & 33.641 & -11.159 & 6.435 & 60.928 \end{bmatrix}$$

While these covariance matrices are by no means identical to each other, they do have similar patterns of small and large, positive and negative values. The variables involved are not so far from being Normally distributed, which causes us to worry about carrying out a test of significance such as Wilks' lambda test (see Chapter 3) to assess whether or not they are sufficiently similar to allow us to pool them together. The p-value resulting from this test is 0.252, which is greater than 5% so we conclude that we have insufficient evidence to claim that the covariance matrices are different from each other. We can thus pool them and get the result shown below:

$$S_* = \begin{bmatrix} 152.607 & -2.063 & -25.576 & -25.861 & -38.610 \\ -2.063 & 296.514 & -5.780 & 2.483 & 12.103 \\ -25.576 & -5.780 & 82.746 & 18.765 & -26.375 \\ -25.861 & 2.483 & 18.765 & 22.200 & 9.841 \\ -38.610 & 12.103 & -26.375 & 9.841 & 77.160 \end{bmatrix}$$

I am now going to use some technical terms that are relevant when dealing with matrices and vectors. The linear discriminant functions, $\underline{m}_i^T S_*^{-1} \underline{x} - \tfrac{1}{2} \underline{m}_i^T S_*^{-1} \underline{m}_i$, are made up of these so they cannot be avoided. However, please do not get scared off by the use of terms such as *invert*, *pre-multiplied, post-multiplied* and *transpose*. It is not necessary to know the details of these mathematical procedures in order to understand the basics of what is going on.

To get the linear discriminant functions $\underline{m}_i^T S_*^{-1} \underline{x} - \tfrac{1}{2} \underline{m}_i^T S_*^{-1} \underline{m}_i$, we need to *invert* this pooled covariance matrix. This inverse of the pooled covariance matrix is *pre-multiplied* by the *transpose* of the vector of means for each of the smoking history groups. To get the second term of the linear discriminant functions, this result is then *post-multiplied* by the vector of means for each of the smoking history groups and this is then multiplied by.

I will now illustrate what results from the previous paragraph using just one of the groups: "never smoked". The procedures for the other groups would be exactly the same, just using the vector of means appropriate to the group in question.

For the "never smoked" group, the vector of means for the variables age, weight, systolic blood pressure, diastolic blood pressure and pulse rate is as follows:

$$\bar{x}_{\text{never smoked}} = \begin{pmatrix} 33.75 \\ 70.50 \\ 107.15 \\ 59.55 \\ 65.30 \end{pmatrix}$$

To get the first half of the linear discriminant functions (the $m_i^T S_*^{-1} x$), we do the inverting and pre-multiplying specified above to get the $m_i^T S_*^{-1}$ and then *post-multiply* by x. Now this vector x contains the actual values of the variables for a particular case. Rather than discuss one particular case right now, let us just use x_1, x_2, x_3, x_4 and x_5 to represent the values of age, weight, systolic blood pressure, diastolic blood pressure and pulse rate. The $m_i^T S_*^{-1}$ and x are then as follows:

$$m_i^T S_*^{-1} = \begin{pmatrix} 1.327 & 0.192 & 1.945 & 1.708 & 1.927 \end{pmatrix}$$

$$x = \begin{pmatrix} x_1 \\ x_2 \\ x_3 \\ x_4 \\ x_5 \end{pmatrix}$$

If you are not familiar with vectors and matrices, please do not worry as to why the $m_i^T S_*^{-1}$ is written as a row and the x as a column. The important thing is to multiply these two vectors together, and the result is as follows:

$$m_i^T S_*^{-1} x = 1.327x_1 + 0.192x_2 + 1.945x_3 + 1.708x_4 + 1.927x_5$$

To get the second half of the linear discriminant functions (the $-\frac{1}{2} m_i^T S_*^{-1} m_i$), we do the inverting, pre-multiplying and post-multiplying specified above, and obtain the figure -247.159. This means that the final linear discriminant function (ldf) for the group "never smoked" is as follows:

$$ldf_{\text{never smoked}} = -247.159 + 1.327x_1 + 0.192x_2 + 1.945x_3 + 1.708x_4 + 1.927x_5$$

For the other smoking history groups, we get the following linear discriminant functions:

$$ldf_{\text{occasional smoker}} = -285.401 + 1.462x_1 + 0.182x_2 + 2.040x_3 + 1.877x_4 + 2.118x_5$$

$$ldf_{\text{ex-smoker}} = -300.280 + 1.538x_1 + 0.210x_2 + 2.053x_3 + 2.017x_4 + 2.104x_5$$

$$ldf_{\text{current smoker}} = -338.679 + 1.656x_1 + 0.182x_2 + 2.165x_3 + 2.223x_4 + 2.211x_5$$

6.5.2 Allocating Cases to the Groups

All this creating of the linear discriminant functions in Section 6.5.1 is rather pointless if we do not then go on to use them. One way of using them to get an idea of which variables discriminate between the groups is discussed in Section 6.6. Here, we are going to concentrate on using them to decide to which group cases should be allocated.

Let us take as an example Case 1 from the dataset of Chapter 1. It has values for age (x_1) of 21, weight (x_2) of 51.1, systolic blood pressure (x_3) of 94, diastolic blood pressure (x_4) of 63 and pulse rate (x_5) of 82. If we put these values of x_1 to x_5 into each of the linear discriminant functions for the four smoking history groups which we created in Section 6.5.1, we get the following:

$$ldf_{\text{never smoked}} = 239.013$$

$$ldf_{\text{occasional smoker}} = 238.288$$

$$ldf_{\text{ex-smoker}} = 235.377$$

$$ldf_{\text{current smoker}} = 230.375$$

So what does this mean? Which of the four groups is Case 1 nearest? Let us return to Equation 6.1 and examine the transformation of the Mahalanobis distance into the linear discriminant functions. This shows that the linear discriminant functions (the bit in the square brackets in the bottom line) is actually doubled and *taken away* from $\underline{x}^T S_*^{-1} \underline{x}$ to create the Mahalanobis distance. Thus, *larger* linear discriminant functions make *smaller* distances. So, looking at the linear discriminant functions for Case 1, we see that the largest is for the "never smoked" group, which means that Case 1 is nearest to the "never smoked" group according to the Mahalanobis distance. We would thus allocate Case 1 to this group on the basis of these linear discriminant functions.

Of course, for Case 1 we know what smoking history group he or she really belongs to, and it is in fact the "never smoked" group. The discriminant analysis has therefore successfully allocated Case 1 to the correct group. If we do this same allocating to a group for all 100 cases in the dataset of Chapter 1, we find that we end up allocating 73 correctly and 27 incorrectly: a 73% success rate. This shows us that although the variables age, weight, systolic blood pressure, diastolic blood pressure and pulse rate are reasonably good at discriminating between the four smoking history groups, they do not give perfect results.

6.6 WHICH VARIABLES DISCRIMINATE BETWEEN GROUPS?

In Section 6.5.2 we found that the variables age, weight, systolic blood pressure, diastolic blood pressure and pulse rate managed to join together in linear discriminant functions to correctly allocate 73% of the cases in the dataset of Chapter 1. But are they all useful, or only some of them?

To answer this question, we can examine the following linear discriminant functions created in Section 6.5.1.

$$ldf_{\text{never smoked}} = -247.159 + 1.327x_1 + 0.192x_2 + 1.945x_3 + 1.708x_4 + 1.927x_5$$

$$ldf_{\text{occasional smoker}} = -285.401 + 1.462x_1 + 0.182x_2 + 2.040x_3 + 1.877x_4 + 2.118x_5$$

$$ldf_{\text{ex-smoker}} = -300.280 + 1.538x_1 + 0.210x_2 + 2.053x_3 + 2.017x_4 + 2.104x_5$$

$$ldf_{\text{current smoker}} = -338.679 + 1.656x_1 + 0.182x_2 + 2.165x_3 + 2.223x_4 + 2.211x_5$$

If one of the variables was completely useless in discriminating between the groups, then we would expect their impact in the linear discriminant functions to be very low compared with the other variables. Here we see that the coefficients in the linear discriminant functions for x_2 (weight) are, in fact, quite low, whereas the other variables have coefficients that are quite similar in size. However, before we go off and draw conclusions directly, we must remember that these coefficients are being multiplied by the values of the variables, and thus the units in which the variables are measured are important. That is, if we have a variable which has very large numbers (for instance, if we measured weight in grammes), then it could have very small coefficients in the linear discriminant functions while still being useful at discriminating between the groups. Similarly, if we had a variable with very small numbers (such as measuring pulse rate in beats per second), then it could have relatively large coefficients in the linear discriminant functions and still not be all that useful at discriminating between the groups.

To overcome this problem, what we need to look at are the linear discriminant function coefficients that would result from using standardised data. If we standardised each of the variables age, weight, systolic blood pressure, diastolic blood pressure and pulse rate (by deducting their means and dividing by their standard deviations), then the linear discriminant functions that would be calculated are as follows:

$$ldf_{\text{never smoked}} = -1.551 - 2.114x_1 + 0.007x_2 - 0.907x_3 - 1.283x_4 - 1.269x_5$$

$$ldf_{\text{occasional smoker}} = -0.130 - 0.113x_1 - 0.161x_2 + 0.054x_3 - 0.261x_4 + 0.533x_5$$

$$ldf_{\text{ex-smoker}} = -0.345 + 1.015x_1 + 0.322x_2 + 0.187x_3 + 0.582x_4 + 0.406x_5$$

$$ldf_{\text{current smoker}} = -2.746 + 2.753x_1 - 0.172x_2 + 1.320x_3 + 1.828x_4 + 1.421x_5$$

Now we can see that all of x_1 (age), x_3 (systolic blood pressure), x_4 (diastolic blood pressure) and x_5 (pulse rate) have relatively large coefficients for at least one or more of the linear discriminant functions. However, x_2 (weight) has small coefficients in all the linear discriminant functions. We can thus conclude that weight is not a good variable at discriminating between the smoking history groups, but the other variables are.

6.7 HOW ACCURATE ARE THE ALLOCATIONS?

In Section 6.5.2 we allocated Case 1 to the "never smoked" group because this group had the largest value for the linear discriminant function, and thus Case 1 was nearer to the "never smoked" group than to any of the other groups (on the basis of the Mahalanobis distance). However, we could not be sure that we were making the correct allocation to the group. Were we as much as 90% sure that Case 1 belonged to the "never smoked" group? Or was it that we were very uncertain and we allocated Case 1 to the "never smoked" group simply because it was marginally more likely than any other group? If we are able to make certain assumptions (see Section 6.7.1), then we can calculate the probability of a case belonging to each of the groups.

6.7.1 Assumptions

In order for the calculations of the probabilities to be believable, we have to make some assumptions about the data. If the assumptions do not hold, then we should not take any notice of the probabilities calculated. This does not mean that the whole of the discriminant analysis is invalid as we have not made any assumptions in previous sections that are absolutely required. There are some that are useful for the linear discriminant function method to perform to its full potential (independence and that the group covariance matrices are similar); but if they are not satisfied, then we do not have to scrap the whole analysis. However, the assumptions required for the probability calculations are as follows. Actually the first three assumptions are the same as are required when undertaking a hypothesis test to see whether or not the mean vectors for the groups involved are the same (see Chapter 3). These assumptions are discussed in more detail in that chapter, so we just concentrate on the fourth assumption here.

1. The cases in the data are independent of each other.
2. The data come from a multivariate normal distribution.
3. The covariance matrices for the groups being investigated are the same.
4. The *prior probability* of a case belonging to a group is the same for all groups.

The last assumption means that before we examine the data for a particular case, we have no idea what group the case may belong to. Using the smoking histories as an example, we are saying that before we looked at the data for Case 1 in Section 6.5.2, it was equally likely that Case 1 belonged to the "never smoked", "occasional smoker", "ex-smoker" or "current smoker" group. Sometimes this may not be a very realistic assumption to make. For instance, in the dataset of Chapter 1, 40% of the cases are in the group "never smoked", with 16% in "occasional smoker", 20% in "ex-smoker" and 24% in "current smoker". On the basis of these figures, we would probably guess that Case 1 is more likely to be in the "never smoked" group. If this assumption of equal prior probabilities cannot be sustained, then there are some methods of calculating probabilities which can overcome this. These are more complicated to calculate and are outside the scope of this book.

6.7.2 Probabilities

To calculate the probabilities of a case belonging to the different groups, we need to have not just the values of the linear discriminant functions for the case, but also the actual Mahalanobis distance between the case and the group. However, thanks to working out Equation 6.1, we can calculate these quite easily. To get the Mahalanobis distance from the linear discriminant function, the last line of Equation 6.1 shows us that we need to calculate $\underline{x}^T S_*^{-1} \underline{x}$ and then take away twice the value of the linear discriminant function. For Case 1, we find that $\underline{x}^T S_*^{-1} \underline{x} = 485.303$. I deliberately leave out the details of how we get this but it is just the same sort of thing that we went through to get the $\underline{m}_i^T S_*^{-1} \underline{x}$ and $-\frac{1}{2} \underline{m}_i^T S_*^{-1} \underline{m}_i$ in Section 6.5.1. The values of the linear discriminant functions for Case 1 are shown in Section 6.5.2. Taking away twice these values from 485.303 gives the following Mahalanobis distances between Case 1 and the different groups.

$$d^2_{\underline{x}, \text{never smoked}} = 7.277$$

$$d^2_{\underline{x}, \text{occasional smoker}} = 8.727$$

$$d^2_{\underline{x}, \text{ex-smoker}} = 14.548$$

$$d^2_{\underline{x}, \text{current smoker}} = 24.754$$

For reasons connected with the multivariate normal distribution, we now take the exponential of $-1/2$ of each of these distances. For the "never smoked", we obtain $e^{-0.5 \times 7.277} = 0.0263$; for the "occasional smoker" group, we obtain 0.0127; for the "ex-smoker" group, we obtain 0.000693; and for the "current smoker" group, we obtain 0.00000422.

The probabilities are then worked out for the "never smoked" group by dividing its 0.0263 by the sum of these values for all the groups. This is $0.0263/(0.02623 + 0.0127 + 0.000693 + 0.00000422)$, which comes to 0.662. Thus, when we allocated Case 1 to the "never smoked" group, there was a 66.2% chance of this being correct (if the assumptions of Section 6.7.1 are correct). We can get probabilities for the other groups as well. For the "occasional smoker" group, the calculations give $0.0127/(0.0263 + 0.0127 + 0.000693 + 0.00000422) = 0.321$. For the "ex-smoker" group, the probability is 0.017; and for the "current smoker" group, the probability is 0.0001.

6.8 TESTING A DISCRIMINANT ANALYSIS

In Section 6.5.2 we said that the variables age, weight, systolic blood pressure, diastolic blood pressure and pulse rate together had a 73% success rate when allocating the cases of the dataset in Chapter 1 to the smoking history groups.

However, when we allocated Case 1 to the "never smoked" group, it could be considered that we were cheating a bit. We decided that the nearest group to Case 1 was "never smoked" on the basis that it had the smallest Mahalanobis distance (and equivalently the largest linear discriminant function value). But as Case 1 did belong to the "never smoked" group, it actually contributed to the calculation of the mean vector for the "never smoked" group. It is thus not too surprising that we decided Case 1 was near the "never smoked" group because it contributed to how the "never smoked" group was defined. Of course, Case 1 was only one of forty cases making up the "never smoked" group, so its influence on what the whole group looked like could be considered relatively small. However, for smaller groups, such as "occasional smokers", each of the sixteen cases does have a relatively important impact on how what characteristics the group has.

Thus, the 73% success rate we observed in Section 6.5.2 may be a bit of an overestimate of how well the linear discriminant functions might do when faced with the task of trying to allocate completely new cases to one of the four smoking history groups. The next two sections suggest ways in which the discriminant analysis may be modified so as to avoid this overestimating of the allocation success rate.

6.8.1 Splitting the Dataset

If you have a sufficiently large number of cases in your dataset, you might consider splitting it into two parts. You could use one part of the dataset to come up with the linear discriminant functions, and then see what happens when these functions are applied to the other part of the dataset. This is likely to give you a better idea of how the linear discriminant functions will perform when allocating new cases that come along.

6.8.2 Cross-Validation

An alternative to splitting the dataset is to perform the discriminant analysis many times. On each occasion, one of the cases in the dataset is left out when the calculations are done to create the linear discriminant functions. This one case is then allocated to a group on the basis of the functions which it had no part in creating. If this procedure is done so that each case in the dataset has been left out on one occasion, then at the end of the process you have a good idea of how often the linear discriminant functions can correctly allocate new cases.

The main problem with this method is that it is very computer intensive. For our dataset with 100 cases, we would, in effect, be doing 100 separate analyses. While this may be possible if you are prepared to wait long enough for your computer to do all the work, sometimes it will not be a practical way of operating. However, if your dataset is large enough to give you problems with how long it would take to operate this procedure, then it may well be large enough to use the method of splitting the dataset discussed in Section 6.8.1.

6.9 OTHER METHODS OF DISCRIMINANT ANALYSIS

There are methods of carrying out a discriminant analysis other than using linear discriminant functions. A couple of methods are discussed below, but readers wanting to know more could look at Tabachnick and Fidell (2013).

6.9.1 Canonical Discrimination

In canonical discrimination for just two groups, a "linear discriminant equation" is created using the p variables chosen. It looks like a regression equation: $y = \beta_0 + \beta_1 x_1 + \beta_2 x_2 + \vdots + \beta_p x_p$. The β parameters are estimated so that the two groups contain y-values that are as different as possible. Another way of thinking of it is that we are undertaking a one-way analysis of variance with the response variable being the y-values, and the groups being the factors. We want to choose values of β to maximise the difference in y-values between the two groups, and so we maximise Between Groups SS/Within Groups SS. To do this maximising, we find the eigenvalues and eigenvectors of $\mathbf{W}^{-1}\mathbf{B}$, where $\mathbf{B}$ is the matrix of Between Groups SS and $\mathbf{W}$ is the matrix of Within Groups SS.

With one linear discriminant equation, y has been created so that the groups are as separated out as possible in a one-dimensional manner, as in Figure 6.1.

If we have more than two groups in the discriminant analysis, then it is possible to define multiple discriminant equations, uncorrelated with each other. For example, two equations separate the groups in two dimensions in Figure 6.2. In general, for k groups, $k - 1$ equations are possible.

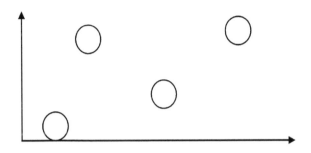

FIGURE 6.1   Groups separated in one dimension.

FIGURE 6.2   Groups separated in two dimensions.

6.9.2 Stepwise Discrimination

In canonical discrimination, the linear discriminant functions that were used looked like regression equations. If the independence and multivariate normality assumptions mentioned in Section 6.7.1 are reasonable, then a series of discriminant analyses can be undertaken to come up with a "best fit" model, just as is done in stepwise multiple regression. That is, variables are added to the discriminant functions (and/or taken out), until it is found (by carrying out significance tests) that adding extra variables or taking any away does not give significantly better discrimination between the groups.

6.10 A STEP-BY-STEP GUIDE TO DISCRIMINANT ANALYSIS USING THE EXCEL ADD-IN

1. You must have a column in Excel that contains the names by which your cases are known. These are called the "case identifiers". They may be names or codes that you can use to identify the different cases, or may be simply case numbers (e.g. case 1, case 2, etc.). You

must also have columns of data containing the variables which you want to use in the cluster analysis and a column that tells Excel which group each case is in.

2. Go through the multivariate analysis add-in's menus until you get the dialogue box for discriminant analysis.

3. In the "Case identifiers:" box, put the range of cells corresponding to the column in which the case names, labels or whatever (see Step 1) are located.

4. In the "Variables to use in analysis:" box, put the range of cells corresponding to the variables you are using in the analysis.

5. In the "Group identifiers:" box, put the range of cells corresponding to the column that indicates which group each case in the dataset is in.

6. Make sure the Yes/No choice for "Variable names in first line of data?" is appropriate for the ranges you have entered at Steps 3, 4 and 5.

7. Click "OK".

6.11 MORE INFORMATION

In various parts of this chapter, I have pointed you towards other sources if you want to find out more about the intricacies of discriminant analysis. Books that deal well with the subject include Everitt and Dunn (2001) and Manley (2005). Tabachnick and Fidell (2013) go into the topic in more depth.

Multidimensional Scaling

7

7.1 WHY DO I WANT TO DO MULTIDIMENSIONAL SCALING?

Most data that you are likely to have when undertaking a study of some kind are information about individual cases. That is, you have a number of cases (e.g. people, organisations, businesses, objects) and you have a number of variables associated with them which come from a process of measuring, asking, observing, etc.

However, sometimes you may have data which is information about pairs or groups of items. For instance, you may have asked a number of people to compare different brands of soft drink, or you might be asking people how similar they think pairs of nations are in the way they conduct international politics. What you then have is information about the soft drinks or nations themselves, and how similar or dissimilar they are from each other. The people who gave you this information are not the focus of the study you are carrying out, but instead it is these other "units" which are important.

Multidimensional scaling can help you analyse this sort of data. On the one hand, it can be thought of in a way similar to factor analysis (see Chapter 4). In factor analysis, you have a matrix of correlations/covariances which are analysed to create a number of different factors. You then try to interpret the factors. You can use multidimensional scaling to do a similar thing, although instead of starting with a correlation or covariance matrix, you start with a different sort of matrix which shows the similarities or dissimilarities. You need to decide how many "dimensions" adequately represent the similarities/ dissimilarities in the same way that in factor analysis you need to decide how many factors adequately represent the correlation/covariance matrix. These "dimensions" have different values for each of the items of interest (e.g. brands of soft drink, nations) which can be examined in a similar way to the loadings produced by a factor analysis.

However, sometimes the object of multidimensional scaling is not to necessarily go as far as trying to interpret the dimensions, but instead it is to try to achieve a picture of the similarities/dissimilarities. This is usually a scatterplot of the first two dimensions, and is thus two-dimensional. If these first two dimensions do quite well in explaining the similarities/dissimilarities, then you have a good "picture" of the data. However, if you need more than two dimensions to get an adequate representation of the similarities/dissimilarities, then this aim is difficult to achieve with any satisfaction. Three dimensions might just be possible if you can use software to spin the three-dimensional plot on a two-dimensional screen, but four or more dimensions are impossible to deal with in any straightforward way.

7.2 WHAT DATA DO I NEED FOR MULTIDIMENSIONAL SCALING?

As mentioned in Section 7.1, ultimately what you need for multidimensional scaling is a matrix of similarities or dissimilarities. However, it is often the case that this matrix is created from variables collected from people, organisations, etc. In the situation where people are being asked about how different nations conduct international politics, they may have been asked to name which two nations were most similar. So, if there were only three nations being asked about, say the United Kingdom (U.K.), United States (U.S.A.), and Germany, then each respondent would either reply U.K./U.S.A. or U.K./Germany or U.S.A./Germany. Once data from a number of respondents had been collected, we could simply count the number of times that each of these three possible responses had been given. This would then have given us a matrix of similarities, with high numbers where lots of people thought the countries were similar, and low numbers where few people thought they were similar.

Alternatively, rather than asking people which pair of countries they think are most similar, a different approach would be to ask for a score to be given to each pair of countries. Low scores could mean that it is thought that nations have similar approaches to international politics and high scores mean that it is thought they have different approaches. If the scores for each of the pairs of countries are averaged over all the respondents, then the resulting averages would be entries in a dissimilarity matrix.

Another situation in which a matrix of similarities/dissimilarities can be created is where a person (or other sort of respondent) has ranked a number of items. Thus, to use the soft drinks brand example again, the respondents were asked to put the brands in order of how much they like the taste. For every possible pair of brands, the difference in the rankings can be calculated. Thus, if there are four brands, A, B, C and D that have been given the ordering 1st = B, 2nd = D, 3rd = C, 4th = A, then A/B are three places apart, A/C are one place apart, A/D are two places apart, B/C are two places apart, B/D are one place apart and C/D are one place apart. This process is repeated for each of the respondents in the study, and the distances between the pairs of brands averaged over all of them. The result is a matrix of dissimilarities, with brands which are consistently ranked similarly having small dissimilarities and brands which are consistently ranked differently having large dissimilarities. It is also possible to carry out the same sort of process when instead of ranks being defined for the brands, scores out of 10 or something similar are awarded for each brand.

In the topic of cluster analysis (see Chapter 5), one of the first steps in hierarchical clustering is to construct a distance matrix according to some measurement criteria. This distance matrix could be used as the dataset for a multidimensional scaling. In this situation, the multidimensional scaling would be used to try and obtain a good two-dimensional picture of the dataset. The idea of interpreting the dimensions in a similar way to the factors in factor analysis is unlikely to be very illuminating as in multidimensional scaling, each case in the dataset would have a loading whereas in factor analysis the loadings would be attached to the variables which makes interpreting them easier. However, it is not impossible that studies might exist where a multidimensional scaling of a cluster analysis type distance matrix could give meaningful dimensions.

It is possible that the data you obtain does not need to be converted into a matrix of similarities or dissimilarities because it is already made up of distances. Consider the situation where you have the time it takes to travel between various towns and cities of the U.K. by train. The data are measures like dissimilarities already, and you may then use them straightaway in a multidimensional scaling. The resulting analysis may give you a two-dimensional picture of the U.K. according to train times. This would have the effect of squashing the U.K. in a north/south direction as fast trains cover these routes, but may also spread out places that are more awkward to get to by train, such as the west coast of Wales and the east coast of Norfolk.

Above, I have been trying to consistently mention that we need a "matrix of similarities/dissimilarities". In essence, this is true but in practice, methods of undertaking multidimensional scaling need the matrix to be dissimilarities. However, if you have a matrix of similarities, it is usually quite straightforward to convert it into a matrix of dissimilarities. Usually, you only have a matrix of similarities if you are counting how many times people are choosing a pair of items, or you have a scoring system where high scores are used to indicate that items are similar to each other. In the first case, similarities can be converted to dissimilarities by counting how many people do not choose a pair of items. For instance, if 89 out of 100 people have judged two items to be similar to each other, then this is equivalent to 11 out of 100 judging them not to be similar. We then get a situation where small numbers are associated with pairs of items that are judged to be similar, and large numbers are associated with pairs of items that are not judged to be similar. We thus have dissimilarities. For the second case, imagine we are using a scoring system of 1 to 10, where 1 indicates that items are dissimilar and 10 indicates that items are similar. We can collect the data from all the respondents and then carry out a conversion to reverse the scale. By calculating 11 minus the score given, we then get 1 indicating items that are similar and 10 indicating items that are dissimilar. The averaged scores are then dissimilarities.

7.3 THE REST OF THIS CHAPTER

Much of this chapter deals with the method of multidimensional scaling called *classical multidimensional scaling*. Section 7.4 deals with this in some detail, but also mentions some difficulties with this method in Section 7.4.1. Methods of multidimensional scaling that help overcome these problems are discussed in Section 7.5. In Section 7.6, a step-by-step guide to using the multidimensional scaling part of the Microsoft Excel add-in is given, and in Section 7.7, sources of information are given for those readers who want to learn more about the topic.

7.4 CLASSICAL MULTIDIMENSIONAL SCALING

The object of multidimensional scaling can be thought of as taking a matrix of dissimilarities and then trying to identify "variables" (or "dimensions") that recreate the dissimilarity matrix when distances between them are calculated.

In classical multidimensional scaling, this is exactly what we do. We start with a distance matrix that we can call **D**. What we want to find is a dataset **X** that is made up of a number of unknown variables. The cases in this dataset will be the items which make up the rows and columns of the distance matrix **D**.

Let us consider the dataset of Chapter 1. In particular, we are going to focus on the responses that the 100 people gave when asked to give scores to pairs of nations according to how similar/dissimilar they thought their foreign policies were. They were asked to give a score between 1 and 10, with 1 indicating very similar and 10 indicating very dissimilar. The average score for each pair of nations over the 100 people form the distance matrix in Table 7.1. It is a dissimilarity matrix rather than a similarity matrix because larger numbers indicate that people thought the nations were more dissimilar in their foreign policies. There is a lot of repetition in the matrix because the distance from, say the U.K. to the U.S.A., is the same as the distance from the U.S.A. to the U.K. Table 7.1 is the matrix **D** for the purposes of this analysis.

So we are looking to create a dataset **X** which would give us the matrix **D** in Table 7.1. If we actually had **X** to start with, then we could calculate **D** straightaway using some distance measure. In classical multidimensional scaling, this distance measure is assumed to be Euclidean distance (see Chapter 5 for further discussion of this measure). However, of course, we do not have **X** but only have **D**. If we can easily go from **X** to **D**, can we easily go from **D** to **X** and thus obtain the dataset we want? Well, it might not be all that easy but it can be done without too much difficulty, as we can see below.

Although we cannot go directly from **D** to **X**, there is a trick we can use. As well as being able to calculate **D** directly from **X**, we can break down the calculations into two parts. The first part would be to create a matrix that is generally called **B** and is equal to $\mathbf{X}^T\mathbf{X}$. This is a bit of matrix multiplication wizardry. If you are not familiar with matrix arithmetic, please do not worry. In effect, all we are doing is squaring the matrix. Squaring is something that is done as part of the calculation of Euclidean distances, so it is not such a weird thing to do.

TABLE 7.1   Distance Matrix for Nations

	U.K.	U.S.A.	FRANCE	GERMANY	RUSSIA	CHINA	AUSTRALIA
U.K.	0.000	2.350	7.600	6.430	7.600	7.520	2.930
U.S.A.	2.350	0.000	7.000	4.000	8.560	8.140	3.190
France	7.600	7.000	0.000	3.870	3.910	4.150	6.130
Germany	6.430	4.000	3.870	0.000	4.020	7.230	3.930
Russia	7.600	8.560	3.910	4.020	0.000	3.180	3.940
China	7.520	8.140	4.150	7.230	3.180	0.000	3.250
Australia	2.930	3.190	6.130	3.930	3.940	3.250	0.000

Once we have this matrix **B**, then the elements of the matrix **D** can be calculated directly from it. Similarly, if we have **D**, we can calculate **B** directly from **D**. Each number in **B** can be calculated using parts of the matrix **D** via the following formula:

$$b_{ij} = -\frac{1}{2}\left\{\delta_{ij}^2 - \delta_{i\bullet}^2 - \delta_{\bullet j}^2 + \delta_{\bullet\bullet}^2\right\}$$

Do not worry about this formula – the important thing is that you know that **B** can be calculated directly from **D**, rather than the details of how it is done. However, because it is really not all that complicated, let me explain the parts of this formula in a bit more detail. The b_{ij} is the number which is on the i-th row and j-th column of the matrix. It does not matter which way around we regard the rows and columns as both **D** and **B** are symmetric. That is, if we consider the line going from the top left corner to the bottom right corner of the matrix, then this is a line of symmetry and values above the line are mirrored by values below the line. Have a look at **D** in Table 7.1. In a similar way, the δ_{ij} is just the part of the **D** matrix in the i-th row and j-th column. You might wonder why I am using δ_{ij} instead of d_{ij} for the elements of **D**. The reason is that in Section 7.5, I will be using both δ_{ij} and d_{ij}, with δ_{ij} being the elements of **D**. It is a convention that is common in multidimensional scaling, and if I were to change it here, then I would run the risk of confusing you if you looked at any other book on the subject.

So, δ_{ij} is one of the observed/calculated distances from the matrix **D**. The δ_{ij}^2 is just the square of the δ_{ij}. The $\delta_{i\bullet}^2$ is calculated by squaring each value in the i-th row of **D** and taking an average of these figures. Similarly, the $\delta_{\bullet j}^2$ is calculated by squaring each value in the j-th column of **D** and taking an average. The $\delta_{\bullet\bullet}^2$ is calculated by squaring all the values in every row and column of **D** and taking an average. So, although it does involve a bit of calculation, the process of creating **B** from **D** only involves relatively simple arithmetic. The matrix **D** of Table 7.1 becomes the matrix **B** shown in Table 7.2.

We have managed to get the matrix $\mathbf{B} = \mathbf{X}^T\mathbf{X}$ from the matrix **D**, but can we get **X** from the matrix **B**? Yes, we can, but unfortunately it is not a case of directly calculating the **X** from **B**. What we need to do is to calculate the *eigenvalues* and *eigenvectors* of **B**. We have done this before for a matrix when doing factor analysis. As in Chapter 4 when discussing factor analysis, I am not going to go into the details of *eigenanalysis* here. What is important for multi-dimensional scaling is that the eigenvectors that result from the eigenanalysis can be thought of as our variables that make up **X**. To be more specific, it is the scaled eigenvectors that make up **X**. The eigenvectors that usually result from an eigenanalysis are in *normalised* form. That is, if you square the values in an eigenvector and add them up, you get exactly one. Scaled eigenvectors are

TABLE 7.2 B Matrix for Nations

	U.K.	U.S.A.	FRANCE	GERMANY	RUSSIA	CHINA	AUSTRALIA
U.K.	19.292	16.010	-12.138	-6.475	-12.403	-10.022	5.738
U.S.A.	16.010	18.250	-8.279	5.676	-20.681	-15.398	4.422
France	-12.138	-8.279	14.192	4.159	6.283	7.092	-11.308
Germany	-6.475	5.676	4.159	9.103	3.302	-12.978	2.786
Russia	-12.403	-20.681	6.283	3.302	13.662	10.382	-0.546
China	-10.022	-15.398	7.092	-12.978	10.382	17.214	3.711
Australia	5.738	4.422	-11.308	-2.786	-0.546	3.711	0.770

TABLE 7.3   Eigenvalues of B Matrix for Nations

	EIGENVALUE	APPROXIMATE AMOUNT OF INFORMATION IN DISTANCE MATRIX REPRESENTED BY EIGENVALUE
1	66.716	51.311%
2	28.451	21.881%
3	11.947	9.189%
4	4.138	3.183%
5	0.000	0.000%
6	−9.198	7.074%
7	−9.573	7.362%

TABLE 7.4   First Two Scaled Eigenvectors of B Matrix for Nations

	EIGENVECTOR	
	1	2
U.K.	3.888	−1.697
U.S.A.	4.443	0.813
France	−2.650	1.973
Germany	0.305	3.289
Russia	−3.594	0.034
China	−3.314	−2.691
Australia	0.921	−1.720

obtained by multiplying each of the values in an eigenvector by the square root of the eigenvalues associated with it. This is exactly what happens in factor analysis when the principal components method is used to extract the factors. The eigenvalues of the **B** matrix in Table 7.2 are shown in Table 7.3 with the first two associated scaled eigenvectors in Table 7.4.

There are two negative eigenvalues shown in Table 7.4. Ideally, there would not be any negative eigenvalues at all. The reason for this is that when you scale the normalised eigenvectors, you need to use the square root of the eigenvalue. If your eigenvalue is negative, then you cannot take the square root without resorting to the mathematical field of imaginary numbers, and this does not really make sense from a statistical point of view.

Negative eigenvalues constitute one of the major problems that classical multidimensional scaling has to face. In most analyses you are going to find that some of the eigenvalues are negative. However, if you only have a small number of negative eigenvalues, and these themselves are not too far from zero, then this issue can be ignored. On the other hand, if you have a lot of negative eigenvalues or there are some which are a long way from zero, then perhaps you should abandon classical multidimensional scaling and use one of the alternative methods of analysis discussed in Section 7.5.

Also shown in Table 7.3 alongside the eigenvalues themselves is the proportion of the information in the distance matrix **D** that the eigenvalues and associated eigenvector represents. This is the same idea as in factor analysis where we can identify the proportion of the variation in the data which each factor can account for. In Table 7.3 we have calculated the proportion by dividing the eigenvalue itself by the sum of all the eigenvalues. Or rather, because we have some negative eigenvalues here, we have used the sum of the absolute values of the eigenvalues (that is, we have ignored the minus signs on the last two eigenvalues). This sum is 130.024 (= 66.716 + 28.451 + 11.947 + 4.138 + 0.000 + 9.198 + 9.573 allowing for rounding). The first eigenvalue is 66.716, so the proportion of the information in the distance matrix which it represents is 66.716/130.024 = 51.311%. Or rather, as we are using a bit of a cheat because of the negative eigenvalues, we should regard this as an approximate figure, as the label for the column in Table 7.3 indicates.

How many of the eigenvalues do we need to worry about? Well, one way of gauging this is to look at a scree plot of the eigenvalues, just as we did in Chapter 4 when considering the number of factors which we wanted to include in a factor analysis. Such a plot is shown in Figure 7.1. From this scree plot we might conclude that the first two eigenvalues are sufficiently large to be of interest, but from eigenvalues number three onwards, they are all quite close to

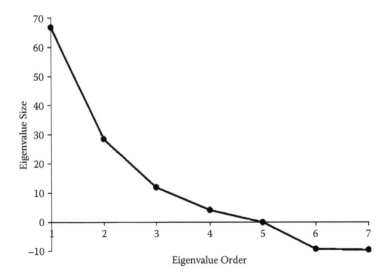

FIGURE 7.1 Scree plot of eigenvalues in Table 7.3.

zero. Another way of deciding how many dimensions to bother with is to look at the cumulative percentage of information in **D** that is represented by the dimensions. I have decided not to include another column in Table 7.3 to show this because with the negative eigenvalues, the percentages already there are only approximate. However, it is easily calculated that the first two dimensions together represent over 70% of the information in the distance matrix. The third dimension adds a bit more but its importance is questionable.

As we have decided that we only need to bother with two dimensions, we can produce a plot to show the nations according to the first two scaled eigenvectors. This plot is shown in Figure 7.2. This shows the relative positions of the seven nations according to these first two, most important, dimensions and shows in one picture the story of the distances of Table 7.1. We see that the U.K. and U.S.A. have been placed near each other and a long way from China, Russia and France. Germany and France have been placed relatively near each other, and Australia is almost midway between the U.K. and China. In the context of this dataset being analysed, being near means that the respondents regard nations as having similar foreign policies and being far apart means the respondents regard them as having dissimilar foreign policies.

We can also try to interpret the scaled eigenvectors of Table 7.4 in the same way we try to interpret factors in factor analysis (see Chapter 4). Here, for the most important eigenvector (eigenvector one), we see that we have large positive values for the U.K. and U.S.A. and large negative values for Russia and China (and to a slightly lesser extent, France). I am no expert on international

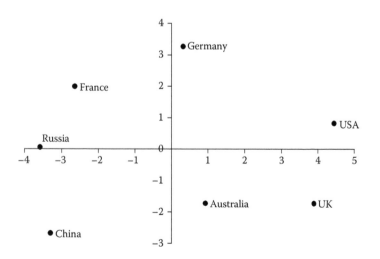

FIGURE 7.2 Plot of first two scaled eigenvectors.

politics but perhaps we could view this underlying factor as people viewing countries' foreign policies as either Anglo-centric or not. The second eigenvector has large positive values for France and Germany and large negative values for the U.K., China and Australia. Could this be seen as people viewing countries' foreign policies as either collegiate or more isolationist?

7.4.1 Problems with Classical Multidimensional Scaling

Having gone to some lengths to show how classical multidimensional scaling works, I must point out that it has fallen out of favour with many analysts. Because it assumes that the distances are Euclidean, it is quite restrictive, and often there is no good reason to make this assumption. Having Euclidean distances implies that the distances are continuous data and can (theoretically) be measured to an infinite degree of accuracy. However, if the distances have been created from a ranking process (e.g. ranking soft drinks in order of preference), then this is not really the case. The analysis can proceed with the false assumption, but the dimensions produced may not be as representative of the true underlying dimensions as would otherwise be the case.

There is also the problem of negative eigenvalues which was mentioned in Section 7.4. If there are a relatively large number of negative eigenvalues or if any of them are large (in a negative direction), then the whole classical multidimensional scaling solution may be called into question.

There are other methods of multidimensional scaling which can get around the problems faced by classical multidimensional scaling. In Section 7.5 we briefly discuss these other methods.

7.5 OTHER METHODS OF MULTIDIMENSIONAL SCALING

As with many multivariate methods of analysis, there is not just one method that can be used to undertake multidimensional scaling. Classical multidimensional scaling (Section 7.4) is a relatively straightforward method from a computational point of view and, because of this, has been widely used in the past. However, modern computing power has opened the possibilities of using other methods which overcome some of the problems associated with classical multidimensional scaling.

The broad aim of multidimensional scaling is to come up with a dataset $\mathbf{X}$ which would recreate the observed distance matrix $\mathbf{D}$ if some sort of distance measure were used. Thus, for each observed element of $\mathbf{D}$, which we will call δ_{ij}, we have a distance, d_{ij} which is calculated from the dataset $\mathbf{X}$. We want these δ_{ij} and d_{ij} to be as close as possible, so we can define the multidimensional scaling problem as one that creates a dataset $\mathbf{X}$ such that the differences between the δ_{ij} and the d_{ij} are as small as possible. To summarise the differences between the δ_{ij} and the d_{ij}, the sum of squared differences between them is frequently used. Although the formula below looks a bit complicated, it is really not that bad. The bit in brackets with the squared sign is simply the difference between the observed distance, δ_{ij} and the distance d_{ij} that comes from the matrix $\mathbf{X}$. This gives us the summary of the differences between the δ_{ij} and the d_{ij}, with squaring taking place so that negative differences do not cancel out positive differences. The Σ symbols are simply telling us to add up all these squared distances.

$$\text{Sum of Squared Distances} = \sum_{i=1}^{n-1} \sum_{j=i+1}^{n} \left(\delta_{ij} - d_{ij} \right)^2$$

In the formula above, only part of the distance matrix $\mathbf{D}$, and the associated matrix of distances estimated from $\mathbf{X}$, is used. This is because in the distance matrix, elements below the *main diagonal* which runs from the top left corner of the matrix to the bottom right corner are repeated above the main diagonal (see, for example, Table 7.1). We do not want to include distances twice, so we only count values above (or below) the main diagonal. Values on the main diagonal are not included in the calculation either. However, these would all be zero (the distance from a case to itself is always zero), so it makes no difference whether you include or exclude them.

Often, rather than use this sum of squared distances, a measure called *stress*, standing for standardised residual *s*um of *s*quares is used. This standardises the sum of squared distances by dividing by the sum of the squared observed distances, and takes the square root:

$$\text{Stress} = \sqrt{\frac{\displaystyle\sum_{i=1}^{n-1} \sum_{j=i+1}^{n} \left(\delta_{ij} - d_{ij} \right)^2}{\displaystyle\sum_{i=1}^{n-1} \sum_{j=i+1}^{n} \delta_{ij}^2}}$$

A mathematical algorithm is used to minimise the value of the stress by finding a dataset $\mathbf{X}$. There are many different algorithms available. The choice of algorithm depends on what you believe about the nature of the distances.

If they are measured on the ratio level of measurement (the size of differences between distances have a concrete meaning and it makes sense to talk about one distance being some multiple of another), then you might use one of a number of algorithms. If they are measured on the interval level of measurement (the size of differences between distances have a concrete meaning but it does not make sense to talk about one distance being some multiple of another – differences in temperature are an example of this because, for example, 4°C cannot be said to be twice 2°C), then you have a choice of another set of algorithms. If the distances result from some ranking or ordering of items, then the distances can be thought of as ordinal (they indicate the order in which items are placed rather than actual distances) and yet another set of algorithms may be used. More information about some of the various algorithms available can be found in Lattin et al. (2003).

Whatever algorithm is chosen, the aim is to minimise the stress. Once this is finally accomplished by the algorithm, you will have a set of dimensions, just as with classical multidimensional scaling. In addition, you can assess how good the created dataset $\mathbf{X}$ is at recreating the original distance matrix $\mathbf{D}$ by looking at the final, minimised, value for the stress. As a rule of thumb, if the stress is as much as 20%, then this is fairly poor; if it gets down to about 10%, then this is not too bad. If it gets down as far as 5%, then you have a very good solution. The ultimate would be a stress of 0%, which would indicate that the dataset $\mathbf{X}$ can exactly recreate the original distances in $\mathbf{D}$.

7.6 A STEP-BY-STEP GUIDE TO MULTIDIMENSIONAL SCALING USING THE EXCEL ADD-IN

1. You must have a distance matrix in Excel. This matrix must have zeros or blanks on the main diagonal (top left cell to bottom right cell) which indicate the distances between an item and itself. It must also have distances between all other pairs of items. These can be in the upper triangle (the area above the main diagonal), or the lower triangle (the area below the main diagonal) or both. If both are given, then they must match each other. Distances must not be negative.

2. Your distance matrix may also have labels in a top row and left-hand column that give the names of the items which form the distance matrix. If given, these must be in the same order for the rows and

columns. If not given, labels will be created for the items, and it will be assumed that the ordering of items is the same in the rows as in the columns.

3. Go through the multivariate analysis add-in's menus until you get the dialogue box for multidimensional scaling.
4. In the "Distance matrix" box, put the range of cells corresponding to the distance matrix you wish to analyse.
5. Make sure the Yes/No choice for "Item names in first row and column of matrix?" is appropriate for the range you have entered at Step 4.
6. Click "OK".

7.7 MORE INFORMATION

Many modern books that deal with multidimensional scaling will completely ignore classical multidimensional scaling because it is not used as commonly as it used to be. However, if you want to know more about classical multidimensional scaling, I suggest you look at Everitt and Dunn (2001), which also discusses metric and non-metric multidimensional scaling. Other books that will let you know more about non-classical methods of multidimensional scaling are Bartholomew et al. (2008), Lattin et al. (2003) and Manly (2005).

Correspondence Analysis

8

8.1 WHY DO I WANT TO DO CORRESPONDENCE ANALYSIS?

All the topics discussed in other chapters of this book have been dealing with data which is continuous, or at least can be treated as continuous. I have not discussed any methods for categorical data other than by having to, first of all, treat them as continuous by allocating some sort of scoring system to the categories (see Chapter 1). In correspondence analysis, we fill this gap to some extent.

For categorical data, the typical method of analysis used time and time again is to produce tables. For two variables, one variable's categories make up the rows and the other variable's categories make up the columns (e.g. Table 8.1). For more than two variables, rows and columns may be made up of various combinations of the categories of the different variables (e.g. Table 8.2).

These tables show the information clearly but it is sometimes difficult to see patterns that exist. The human eye is much better at assessing pictures than figures, so it would be nice to have graphical versions of tables to look at. This is what correspondence analysis creates for us.

TABLE 8.1   Counts for Smoking History and Gender

| | | GENDER | | TOTAL |
		MALE	FEMALE	
Smoking	Never smoked	15	25	40
History	Occasional smoker	11	5	16
	Ex-smoker	9	11	20
	Current smoker	13	11	24
	Total	48	52	100

8.2 WHAT DATA DO I NEED FOR CORRESPONDENCE ANALYSIS?

For correspondence analysis, we need categorical data. These may be variables which have ordered categories (e.g. the age-groups in Table 8.2) or unordered categories (e.g. gender). Correspondence analysis can be carried out with two categorical variables, or more than two. Continuous data can be used, but first it must be grouped into categories. This can sometimes be done by examining a histogram of the continuous variable and seeing where it makes sense to put breaks between categories.

8.3 THE REST OF THIS CHAPTER

Although correspondence analysis is a multivariate technique, much of this chapter focusses on methods for dealing with just two variables. This is because the ideas behind the technique can be most easily explained in this context, and the method which we discuss here for dealing with more than two variables is a relatively straightforward extension of the two-variable ideas. In Section 8.4 we start our exploration of correspondence analysis by going back to the basic chi-square test of independence which is carried out so often on tables of categorical data. We end up obtaining a graphical display of the information in the table. In Section 8.5 we consider how this graphical display may be obtained when we have variables with numerous categories. Section 8.6 deals with the various forms of plots that can be obtained, and we eventually reach a truly multivariate scenario in Section 8.7 when

TABLE 8.2 Counts for Gender, Age Group and Smoking History

Gender				NEVER SMOKED	OCCASIONAL SMOKER	EX-SMOKER	CURRENT SMOKER	TOTAL
						SMOKING HISTORY		
Male	Age Group	18–29		6	2	1	1	10
		30–39		3	4	2	1	10
		40–49		3	4	3	3	13
		50–59		1	0	2	5	8
		60+		2	1	1	3	7
Female	Age group	18–29		11	1	1	0	13
		30–39		8	1	1	1	11
		40–49		5	2	2	1	10
		50–59		1	0	3	3	7
		60+		0	1	4	6	11
		Total		40	16	20	24	100

we consider correspondence analysis of more than two variables. We conclude with a step-by-step guide to undertaking correspondence analysis using this book's Microsoft Excel add-in in Section 8.8 and give further sources of information in Section 8.9.

8.4 CHI-SQUARE DISTANCES, INERTIA AND PLOTS

8.4.1 A Chi-Square Test of Independence

Let us consider the data for smoking history and gender in Table 8.1. A statistical analysis that is often carried out on such data is a chi-square test of independence. The null hypothesis for this test is that there is no association between smoking history and gender. The alternative is that there is an association. The chi-square test of independence is a very well-known technique, so details will not be given here. However, the numbers of "expected" cases that are used in the necessary calculations are shown in Table 8.3. None of the "expected" numbers are less than 5, so we can be happy that the chi-square approximation is appropriate for this analysis.

Combining the observed numbers in Table 8.1 and expected numbers in Table 8.3 in the usual way for the chi-square test of independence, we obtain a chi-square value of 4.965. The number of degrees of freedom associated with this test are the number of smoking history categories minus one, multiplied by the number of gender categories minus one. This is 3×1 which is, of course, 3. The p-value that results from a chi-square value of 4.965 with 3 degrees of freedom is 0.174. As this is more than 0.05, we conclude that

TABLE 8.3   "Expected" Counts for Smoking History and Gender

		GENDER		
		MALE	FEMALE	TOTAL
Smoking History	Never smoked	19.20	20.80	40
	Occasional smoker	7.68	8.32	16
	Ex-smoker	9.60	10.40	20
	Current smoker	11.52	12.48	24
	Total	48	52	100

we have insufficient evidence to reject the null hypothesis at the 5% level of significance.

8.4.2 Inertia

The chi-square value of 4.965 from Section 8.4.1 reflects the amount of information in Table 8.1 about the association between age group and gender. In correspondence analysis, the average contribution to this chi-square value by each case in the dataset is called the *inertia*. This is the chi-square value divided by the number of cases, which in this case is 4.965/100 = 0.04965. In Section 8.5, the importance of inertia in correspondence analysis will become clear.

8.4.3 Plotting Chi-Square Distances

How might we go about plotting the information contained in Table 8.1 in a graph? As the counts of people in each cell of the table would change depending on the size of the dataset, perhaps it would be better to consider percentages, as in Tables 8.4 and 8.5. In Table 8.4 we have calculated the percentages so that they relate to the proportion of each gender which is in each smoking history group. In Table 8.5 we have calculated the percentages so that they relate to the proportion of each smoking history group which is male or female.

From Table 8.5 we can construct Figure 8.1 which shows the different smoking history groups plotted according to the percentages that are calculated for male and female.

The first thing that strikes you about Figure 8.1 is that the points are all in a straight line. Is this a coincidence? No it is not. The male percentage and female percentage have to add up to 100%, so that forces the points to lie

TABLE 8.4 Percentages in Each Smoking History Group for Each Gender

| | | GENDER | | |
		MALE	FEMALE	TOTAL
Smoking History	Never smoked	31.25%	48.08%	40.00%
	Occasional smoker	22.92%	9.62%	16.00%
	Ex-smoker	18.75%	21.15%	20.00%
	Current smoker	27.08%	21.15%	24.00%
	Total	100%	100%	100%

TABLE 8.5   Percentages in Each Gender for Each Smoking History Group

| | | GENDER | | |
		MALE	FEMALE	TOTAL
Smoking	Never smoked	37.50%	62.50%	100%
History	Occasional smoker	68.75%	31.25%	100%
	Ex-smoker	45.00%	55.00%	100%
	Current smoker	54.17%	45.83%	100%
	Total	48.00%	52.00%	100%

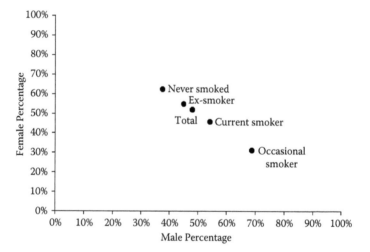

FIGURE 8.1   Gender percentages for smoking history groups.

on a straight line. Looking beyond this, we see that the "Never smoked" and "Ex-smoker" groups lie quite close together, indicating that they have similar male/female percentages, and they are also near the overall "Total" male/female percentage in the dataset. "Occasional smokers" appear to have a quite different male/female split, with males dominating this group to a greater extent than any other group.

However, this section is entitled "Plotting Chi-Square Distances". What we have in Figure 8.1 when we consider how close points are to each other are Euclidean distances between them. We have discussed Euclidean distances elsewhere in this book (Chapter 5). If we consider the Euclidean distance between the closest two categories in Figure 8.1, "Never smoked" and

"Ex-smoker", we get the following. The Euclidean distance is calculated by taking the square of the difference in the male percentages, adding the square of the difference in the female percentages and then taking the square root:

$$\text{Euclidean distance} = \sqrt{(37.50 - 68.75)^2 + (45.00 - 55.00)^2} = 10.607$$

This calculation of the Euclidean distance treats the male percentage and female percentage on the same basis. It takes no account of the fact that, in general, we would expect the female percentages to be (slightly) higher because there are (slightly) more females than males in the dataset. We need to standardise the figures in some way so that female figures are made to be no more important in the calculation than the male figures. The sort of standardising that is done when calculating distances in cluster analysis (see Chapter 5) is not appropriate here as our data are counts of people. Instead, what is done is to divide each of the squared differences in the calculation by the overall percentage in the dataset who are male or female. We thus get a distance which is called a chi-square distance, as follows:

$$\text{Chi-square distance} = \sqrt{\frac{(37.50 - 68.75)^2}{48} + \frac{(45.00 - 55.00)^2}{52}} = 1.501$$

The actual value of the chi-square distance is quite different from the value of the Euclidean distance, but this need not be a concern. They are measuring the distance in different ways, so it is not surprising that they produce quite different figures.

There is another reason why we choose the chi-square distance in preference to the Euclidean distance. This is because of the way in which categories are frequently capable of being combined or divided. Suppose that originally, instead of having just a "female" category, we had a "young female" and an "older female" category. If we created a table showing the percentages of each gender category that was in each smoking history category, we might discover Table 8.6. This indicates that there is no difference between the "young female" and "older female" groups in terms of their smoking history, and we might as well combine the categories, as in Table 8.4.

However, what if we did not combine the categories and instead proceeded to produce Table 8.7 showing the percentage of each gender category which was in each smoking history group? If we were now to calculate the distance from the "Never smoked" category to the "ex-smoker" category, there is no reason for us to expect to get any different result from the calculations above when the "young female" and "older female" groups were combined.

TABLE 8.6   Percentages in Each Smoking History Group for Each Expanded Gender Category

		GENDER			
		MALE	YOUNG FEMALE	OLDER FEMALE	TOTAL
Smoking	Never smoked	31.25%	48.08%	48.08%	40.00%
History	Occasional smoker	22.92%	9.62%	9.62%	16.00%
	Ex-smoker	18.75%	21.15%	21.15%	20.00%
	Current smoker	27.08%	21.15%	21.15%	24.00%
	Total	100.00%	100.00%	100.00%	100.00%

TABLE 8.7   Percentages in Each Expanded Gender Category for Each Smoking History Group

		GENDER			
		MALE	YOUNG FEMALE	OLDER FEMALE	TOTAL
Smoking	Never smoked	37.50%	18.75%	43.75%	100%
History	Occasional smoker	68.75%	9.38%	21.88%	100%
	Ex-smoker	45.00%	16.50%	38.50%	100%
	Current smoker	54.17%	13.75%	32.08%	100%
	Total	48.00%	15.60%	36.40%	100%

There is no difference in the relationships between Table 8.5 and Table 8.7, so why should the differences between the smoking history groups change?

The Euclidean and chi-square distances between the "Never smoked" and "Ex-smoker" categories can be calculated in exactly the same way as for Table 8.5, as follows.

Euclidean distance =

$$\sqrt{(37.50-45.00)^2 + (18.75-16.50)^2 + (43.75-38.50)^2} = 9.427$$

Chi-square distance =

$$\sqrt{\frac{(37.50-45.00)^2}{48} + \frac{(18.75-16.50)^2}{15.6} + \frac{(43.75-38.50)^2}{36.4}} = 1.501$$

TABLE 8.8   Chi-Square Scaled Percentages in Each Gender for Each Smoking History Group

		GENDER	
		MALE	FEMALE
Smoking History	Never smoked	$\dfrac{37.50\%}{\sqrt{48\%}} = 0.541$	$\dfrac{62.50\%}{\sqrt{52\%}} = 0.867$
	Occasional smoker	$\dfrac{68.75\%}{\sqrt{48\%}} = 0.992$	$\dfrac{31.25\%}{\sqrt{52\%}} = 0.433$
	Ex-smoker	$\dfrac{45.00\%}{\sqrt{48\%}} = 0.650$	$\dfrac{55.00\%}{\sqrt{52\%}} = 0.763$
	Current smoker	$\dfrac{54.17\%}{\sqrt{48\%}} = 0.782$	$\dfrac{45.83\%}{\sqrt{52\%}} = 0.636$
	Total	$\dfrac{48.00\%}{\sqrt{48\%}} = 0.693$	$\dfrac{52.00\%}{\sqrt{52\%}} = 0.721$

We see that the Euclidean distance has changed from 10.607 to 9.427 but the chi-square distance has remained the same at 1.501. So, both for the reason of wanting to undertake some standardisation of the percentage differences and to ensure that the way in which we define the categories does not affect the distances, we prefer to use chi-square distances when undertaking correspondence analysis.

However, we live in a world where distances between points on a graph are Euclidean. How can we change Figure 8.1 so that the distances between categories are chi-square distances? The answer is to amend Table 8.5 by dividing each percentage by the square root of the total percentage in the dataset that is male or female, as appropriate. The result is Table 8.8, which can then be displayed in a graph as Figure 8.2. The result is very much like Figure 8.1, so the conclusions drawn from that graph are equally valid here.

A further feature of the chi-square scaled percentages shown in Table 8.8 will come as no surprise if you consider that we are calling them "chi-square" scaled. If we take the squared Euclidean distance from one of the smoking history categories to the "Total", multiply this by the number of people in that smoking history group, and add this to the same quantity for all the other smoking history groups, then we get the original chi-square

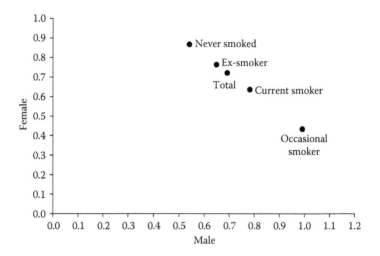

FIGURE 8.2   Chi-square scaled gender percentages for smoking history groups.

value of 4.965 from the test of independence carried out in Section 8.4.1, as shown below.

$$\chi^2 = 40 \times \left[(0.541 - 0.693)^2 + (0.867 - 0.721)^2 \right] + 16 \times \left[(0.992 - 0.693)^2 \right.$$

$$+ (0.433 - 0.721)^2 + 20 \times \left[(0.650 - 0.693)^2 + (0.763 - 0.721)^2 \right]$$

$$+ 24 \times \left[(0.782 - 0.693)^2 + (0.636 - 0.721)^2 \right]$$

$$= 4.965$$

8.4.4 Adding Extreme Categories to the Plots

Figure 8.2 gives us a graphical view of how similar the different smoking history groups are to each other, but does not give us any information about how they relate to gender. To add gender to the graph, we amend Table 8.5 to show two additional categories in the rows: one relating to a hypothetical group where 100% of the people are male, and one relating another hypothetical group where 100% of the people are female. The result is Table 8.9.

TABLE 8.9 Percentages in Each Gender for Each Smoking History Group with Extreme Categories

| | | GENDER | | |
		MALE	FEMALE	TOTAL
Smoking History	Never smoked	37.50%	62.50%	100%
	Occasional smoker	68.75%	31.25%	100%
	Ex-smoker	45.00%	55.00%	100%
	Current smoker	54.17%	45.83%	100%
	Male	100%	0%	100%
	Female	0%	100%	100%
	Total	48.00%	52.00%	100%

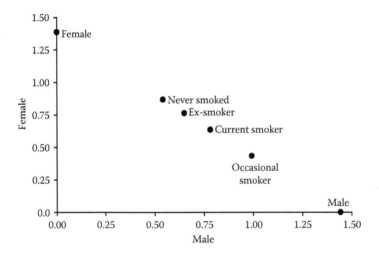

FIGURE 8.3 Chi-square scaled gender percentages for smoking history groups with extreme categories.

If we create a graph of chi-square distances between the row categories like Figure 8.2, we get Figure 8.3. It shows that the "Female" category is the one that is nearest to the "Never smoked" and "Ex-smoker" groups, whereas the "Male" category is nearest to the "Occasional smoker" and "Current smoker" groups. We can thus see in the graph the associations that exist in the data. In this case, the associations are not very strong – the extreme categories in Figure 8.3 are not particularly near any of the smoking history groups, and the chi-square test of independence was not significant at the 5% level of significance.

8.5 MORE DIMENSIONS

In Section 8.4 the graphs that were produced all contained the points for the various smoking history groups (and extreme categories) in a straight line. This means that although the graphs we produced were two dimensional, we could have just had a one-dimensional plot. The distances between the points would have been the same in the one-dimensional plot, so no information would have been lost.

Consider now Table 8.10 which shows a table of smoking history against age group. It is not appropriate to undertake a chi-square test of independence on this table because for many of the cells, the expected count under the null hypothesis of no association is rather small. An "exact test" which tries to conduct the hypothesis test using probability theory runs into problems because of the non-trivial number of cells in the table and the size of the dataset, and so cannot overcome this problem. Combining the 50–59 and 60+ age groups appears to be a possible solution as they have similar profiles over the smoking history categories. However, the resulting table still cannot be analysed appropriately with a chi-square test of independence, and exact tests are still problematic.

To try to examine the relationship between smoking history and age group, let us consider Table 8.11. This is the same sort of table as Table 8.5, showing (in this case) the percentage of each smoking history group that lies in each age group. For Table 8.5, because there were just two column categories, we were able to produce a plot showing the percentages (Figure 8.1). We later amended this plot to show chi-square distances (Figure 8.2) for the reasons explained in Section 8.4.3. However, we now have five column categories in Table 8.11. How many dimensions would we need to plot these? Well, initially

TABLE 8.10   Counts for Smoking History and Age Group

		AGE GROUP					
		18–29	30–39	40–49	50–59	60+	TOTAL
Smoking	Never smoked	17	11	8	2	2	40
History	Occasional smoker	3	5	6	0	2	16
	Ex-smoker	2	3	5	5	5	20
	Current smoker	1	2	4	8	9	24
	Total	23	21	23	15	18	100

TABLE 8.11 Percentages in Each Smoking History Group for Each Age Group

		AGE GROUP					
		18–29	30–39	40–49	50–59	60+	TOTAL
Smoking	Never smoked	42.50%	27.50%	20.00%	5.00%	5.00%	100%
History	Occasional smoker	18.75%	31.25%	37.50%	0.00%	12.50%	100%
	Ex-smoker	10.00%	15.00%	25.00%	25.00%	25.00%	100%
	Current smoker	4.17%	8.33%	16.67%	33.33%	37.50%	100%
	Total	23.00%	21.00%	23.00%	15.00%	18.00%	100%

we might consider that we need five dimensions because we have five categories. However, if we go back to Figure 8.2, we see that although we were plotting in two dimensions, we could have represented the plot in just one dimension as the points were in a straight line. Thus, to completely represent the percentages in Table 8.11, we do not need five dimensions but can get away with one less: four dimensions.

You may be less than impressed with the idea that we "only" need four dimensions to completely represent the percentages in Table 8.11. If so, you are right – using the paper (or perhaps screen) on which you are reading this, we can only represent two dimensions with any satisfaction. We might try and get away with three dimensions with some clever computer-aided spinning of a plot that is meant to be three-dimensional, or (if it was worth the effort) make a hologram. However, attempting to represent four dimensions is not a realistic aim. If we had a variable with even more categories than age group, we would need even more dimensions.

In Section 8.5.1 we try to overcome this problem of living in a world with limited physical dimensions. There is a price to pay for coping with this but the amount of information we lose can be quantified.

8.5.1 Reducing a Plot to Two Dimensions

What we want to do is reduce a plot that would normally take more than two dimensions down to just two dimensions whilst keeping as much of the information about the chi-square distances between the points as possible. In a way, it is similar to the task of multidimensional scaling (see Chapter 7) where we might want to represent a multidimensional solution in just two dimensions. To do this in correspondence analysis, we use a mathematical technique called *singular value decomposition.* You will be grateful to hear that I am not going to go into how this works, but it is a technique that is applied to matrices and has connections with eigenanalysis that we encounter elsewhere in this book (see Chapter 4 on factor analysis, Chapter 6 on discriminant analysis and Chapter 7 on multidimensional scaling). If you want to know more about the singular value decomposition procedure, there are many books and other resources available.

What the singular value decomposition does for us is produce the co-ordinates that we need for the two-dimensional plot. In fact, it will produce the co-ordinates not only for the two dimensions that we want, but also for all the possible dimensions. So, for Table 8.11, will it produce four dimensions for us? No it will not, because mathematically the table of four rows and five columns can, in fact, be represented by only three dimensions. This is because although we can represent five columns with four dimensions, the four rows can be represented with three dimensions. The singular value decomposition is so efficient that it can represent all the information in the table with just these three dimensions.

We will consider the co-ordinates for the two-dimensional plot in Section 8.6 but for now let us still consider the problem that faces us of having a three-dimensional solution that we wish to display in two dimensions. If you have read other chapters of this book which include discussion of the results of an eigenanalysis, you will not be surprised to hear that as well as supplying co-ordinates for three dimensions, the singular value decomposition supplies them in order so that most of the information from the table is contained in the first dimension, followed by the second dimension and then the third dimension. Table 8.12 shows the inertia supplied by each of the dimensions. In total, the inertia is 0.3918, which is the chi-square value (39.18) that would be calculated from Table 8.10, divided by the number of cases (100). We see that the first dimension supplies over 87% of the total inertia and the second dimension over 11%. Between them, these first two dimensions account for almost 99.5% of the total inertia and thus of the information in Table 8.10. A plot of these first two dimensions thus displays

TABLE 8.12   Analysis of Inertia for Table of Smoking History and Age Group

DIMENSION	INERTIA	% OF INERTIA	CUMULATIVE %
1	0.3442	87.857%	87.857%
2	0.0456	11.631%	99.488%
3	0.0020	0.512%	100.000%
Total	0.3918		

almost all the available information. However, we postpone any discussion of this plot until Section 8.6.

8.6 ROW, COLUMN AND SYMMETRIC NORMALISATIONS

Some readers will have noted that in Sections 8.4 and 8.5, I have been dealing with tables by calculating percentages such that the rows add up to 100%. In Section 8.4, this eventually led to a plot (Figure 8.3), where co-ordinates calculated for each smoking history group with extreme points for each gender. However, what if I calculated percentages so that the columns added up to 100% instead? Would I then obtain a different plot where co-ordinates were calculated for each gender, and each smoking history group had an extreme point? The answer is, "Yes, it would".

What we have done in Sections 8.4 and 8.5 is apply what is called a *row normalisation* to the tables so that percentages add up to 100% over each row. We could equally have applied a *column normalisation* to the tables so that the percentages add up to 100% for each column. For Table 8.10, a plot resulting from the row normalisation is shown in Figure 8.4. For the column normalisation, the plot is shown in Figure 8.5. A third type of normalization, called a *symmetric normalization,* can also be produced as a kind of summary of the row and column normalisations. In Figure 8.4 the age groups are represented as extreme categories. In Figure 8.5 they are represented as non-extreme categories. While this is happening, the smoking history groups are represented the other way around. To put them both on the same footing, the extreme and non-extreme points can be averaged to produce Figure 8.6 for the symmetric normalisation. The co-ordinates for the points in Figure 8.6 are obtained by multiplying together the co-ordinates for the extreme and non-extreme points and then taking the square root.

Why do we have three different sorts of normalisation? Well, sometimes it is easier to interpret the relationships between the variables from one plot than from another. It is not unusual for a row normalisation plot or column normalisation plot to produce a bunch of points very close together in the middle, a long way from the extreme points. Looking at the symmetric normalisation plot can often resolve this. In Figures 8.4 through 8.6, we get the same interpretation whichever plot we consider. Being an ex-smoker or current smoker is more closely related to the older age groups (50–59 and 60+), while never smoking is more associated with the youngest age group (18–29).

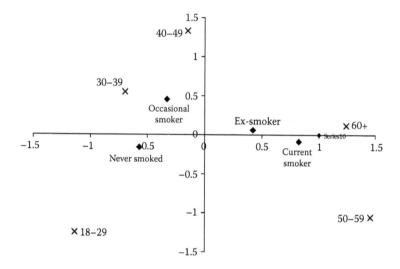

FIGURE 8.4   Row normalisation plot for Table 8.10.

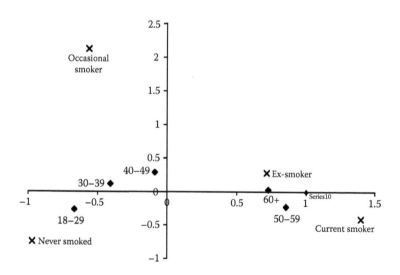

FIGURE 8.5   Column normalisation plot for Table 8.10.

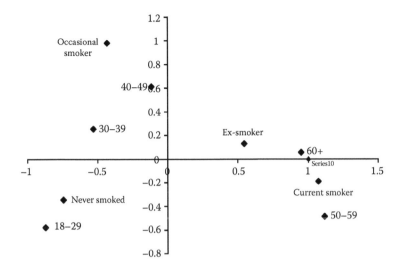

FIGURE 8.6   Symmetric normalisation plot for Table 8.10.

8.7 CORRESPONDENCE ANALYSIS WITH MORE THAN TWO VARIABLES

What we have considered so far in this chapter has been rather restricted because we have only been considering two variables. For this technique to be truly classed as multivariate, it must be able to cope with more than two variables, and indeed it can. However, this *multiple correspondence analysis* does have some shortcomings, as discussed in Section 8.7.1.

8.7.1 The Burt Matrix

To undertake correspondence analysis with more than two variables, a table called a *Burt matrix* is created. For the variables gender, age group and smoking history in the dataset of Chapter 1, this Burt matrix is shown in Table 8.13. It is created by taking all the categories for all the variables being used and using this entire list as the categories in both the rows and columns of a table. In Table 8.13, the first two lines of figures in the table are individual tables of gender against gender, gender against age group and gender against smoking history. The next five lines are individual tables of age group against gender,

TABLE 8.13 Burt Matrix for Gender, Age Group and Smoking History

	FEMALE	MALE	18–29	30–39	50–59	40–49	60+	NEVER SMOKED	CURRENT SMOKER	EX-SMOKER	OCCASIONAL SMOKER
Female	52	0	13	11	7	10	11	25	11	11	5
Male	0	48	10	10	8	13	7	15	13	9	11
18–29	13	10	23	0	0	0	0	17	1	2	3
30–39	11	10	0	21	0	0	0	11	2	3	5
50–59	7	8	0	0	15	0	0	2	8	5	0
40–49	10	13	0	0	0	23	0	8	4	5	6
60+	11	7	0	0	0	0	18	2	9	5	2
Never smoked	25	15	17	11	2	8	2	40	0	0	0
Current smoker	11	13	1	2	8	4	9	0	24	0	0
Ex-smoker	11	9	2	3	5	5	5	0	0	20	0
Occasional smoker	5	11	3	5	0	6	2	0	0	0	16

age group against age group and age group against smoking history. The final four lines are individual tables of smoking history against gender, smoking history against age group and smoking history against smoking history. To analyse this matrix, the same methods as in Sections 8.4 through 8.6 are used, as if we simply had a table of one variable against another.

The Burt matrix contains much of the information that you would want to include in an analysis of the relationships between gender, age group and smoking history. Specifically, it contains information about gender against age group, gender against smoking history and age group against smoking history. However, the Burt matrix contains all this information twice because it has each of these tables twice (that is, it has age group against gender as well as gender against age group, for example). It also has rather a lot of redundant information in it. The individual tables of gender against gender, age group against age group and smoking history against smoking history are not terribly interesting! Because of these issues, correspondence analysis of the Burt matrix is sometimes subject to some criticism. Alternative methods for correspondence analysis of multiple variables have been proposed but they are not pursued here as the Burt matrix approach is still used quite often and the other more advanced methods are beyond the scope of this book.

8.7.2 Analysing the Burt Matrix

When analysing the Burt matrix, the choice of whether to look at the row, column or symmetric normalisation becomes straightforward. As the categories in the rows of Table 8.13 are the same as those in the columns, a plot of either the row or column normalisation would have all the categories in both extreme and non-extreme positions. This makes no sense. It is thus the symmetric normalisation plot that we wish to study. For the Burt matrix of Table 8.13, we obtain the following analysis of inertia (Table 8.14) and plot of the first two dimensions (Figure 8.7).

From Figure 8.7 we see that the relationships between age group and smoking history observed in Section 8.6 can still be seen, with older people more likely to be ex-smokers or current smokers, and younger people more likely to be in the never smoked group. We can also see the relationships between gender and smoking history observed in Section 8.4, with females more likely than males to have never smoked. However, we should note from Table 8.14 that this plot of the first two dimensions only accounts for just over 48% of the information in the Burt matrix. This is not particularly good, so we should be aware that there may be patterns that exist in the data which are not shown in Figure 8.7.

TABLE 8.14   Analysis of Inertia for Burt Matrix of Table 8.13

DIMENSION	INERTIA	% OF INERTIA	CUMULATIVE%
1	0.2821	28.475	28.475
2	0.1958	19.768	48.244
3	0.1246	12.577	60.821
4	0.1203	12.145	72.965
5	0.1111	11.217	84.182
6	0.0805	8.122	92.304
7	0.0578	5.834	98.137
8	0.0185	1.863	100.000
9	0.0000	0.000	100.000
10	0.0000	0.000	100.000
Total	0.9906		

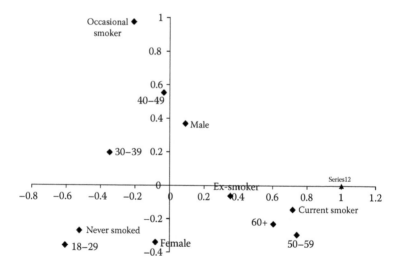

FIGURE 8.7   Symmetric normalisation plot for Burt matrix in Table 8.13.

8.8 A STEP-BY-STEP GUIDE TO CORRESPONDENCE ANALYSIS USING THE EXCEL ADD-IN

1. You must have a column in Excel that contains the names by which your cases are known. These are called the "case identifiers". They may be names or codes that you can use to identify the different cases, or may be simply case numbers (e.g. case 1, case 2, etc.). You must also have columns of data containing the variables which you want to use in the correspondence analysis. These should be categorical variables.

2. Go through the multivariate analysis add-in's menus until you get the dialogue box for correspondence analysis.

3. In the "Case identifiers:" box, put the range of cells corresponding to the column in which the case names, labels or whatever (see Step 1) are located.

4. In the "Variables to use in analysis:" box, put the range of cells corresponding to the variables you are using in the analysis.

5. Make sure the Yes/No choice for "Variable names in first line of data?" is appropriate for the ranges you have entered at Steps 3 and 4.

6. Click "OK".

8.9 MORE INFORMATION

Correspondence analysis is one of the least used of all the multivariate techniques discussed in this book but there is still quite a lot of literature that exists for those readers who want to know more. Books I would recommend that discuss the topic in varying degrees of depth are ones by Bartholomew et al. (2008), Everitt and Dunn (2001), and Manly (2005).

References

Afifi, A., May, S. and Clark, V.A. (2012). *Practical Multivariate Analysis*, 5th edition. Chapman & Hall/CRC: Boca Raton, Florida.

Andrews, D.F. (1972). Plots of high-dimensional data, *Biometrics*, 28, pp. 125–136.

Bartholomew, D.J., Steele, F., Moustaki, I. and Galbraith, J.I. (2008). *Analysis of Multivariate Social Science Data*, 2nd edition. Chapman & Hall/CRC: Boca Raton, Florida.

Brown, B.L., Hendrix, S.B., Hedges, D.W. and Smith, T.B. (2012). *Multivariate Analysis for the Biobehavioral and Social Sciences: A Graphical Approach*. John Wiley & Sons: Hoboken, New Jersey.

Everitt, B.S. and Dunn, G. (2001). *Applied Multivariate Data Analysis*, 2nd edition. John Wiley & Sons Ltd.: Chichester, U.K.

Everitt, B.S., Landau, S., Leese, M. and Stahl, D. (2011). *Cluster Analysis*, 5th edition. John Wiley & Sons Ltd.: Chichester, U.K.

Fabrigar, L.R. and Wegener, D.T. (2012). *Exploratory Factor Analysis*. Oxford University Press: New York.

Field, A. (2009). *Discovering Statistics Using SPSS*, 3rd edition. Sage: London.

Irwing, P., Cammock, T. and Lynn, R. (2001). Some evidence for the existence of a general factor of semantic memory and its components, *Personality and Individual Differences*, 30, pp. 857–871.

Lattin, J., Carroll, J.D. and Green, P.E. (2003). *Analyzing Multivariate Data*. Brooks/Cole Thomson Learning: Belmont, California.

Lynn, R., Irwing, P. and Cammock, T. (2001). Sex differences in general knowledge, *Intelligence*, 30, pp. 27–39.

Manly, B.F.J. (2005). *Multivariate Statistical Methods: A Primer*. Chapman & Hall/CRC: Boca Raton, Florida.

Morrison, D.F. (2005). *Multivariate Statistical Methods*, 4th edition. Brooks/Cole Thomson Learning: Belmont, California.

Tabachnick, B.G. and Fidell, L.S. (2013). *Using Multivariate Statistics*, 6th edition. Pearson Education Boston.

Index